PRAISE FOR *Anne Frank*

"Prose is clear-headed, tough, and fair, and her book, though in places immensely sad, is superb. It should be cherished alongside the masterpiece that inspired it." —*Boston Sunday Globe*

"Prose admirably recreates the events in the attic over the years—no small feat—[with] all the drama of a classic whodunit . . . Transcendent criticism . . . [A] case so brilliantly proven." —*Chicago Tribune*

"Fascinating . . . riveting to read . . . This book will hold your attention in every chapter and will do more than that. I found myself feeling again the tug at the heart, the clamp of fear, the approaching doom. Prose evokes the dread while filling out the story of Anne Frank with many factual details most readers will not have known . . . Readers can be grateful to Francine Prose for the scholarship, clarity and energy with which she has gathered and interpreted the several storms that still surround one girl's life and her book." —Anne Roiphe, *Moment Magazine*

"Francine Prose . . . takes Anne's story and adds to it a new perspective. . . . Prose tells this story with tremendous beauty, pathos and a profound awareness of tragic coincidence." —*San Francisco Chronicle*

"Provocative . . . penetrating . . . With *Anne Frank*, then, Prose means to remove Frank from the wistful amber of her posthumous celebrity and reveal her to us in a more realistic light." —*Los Angeles Times*

"Impassioned . . . compelling . . . No one has made the case as convincingly and forcefully as Francine Prose does that Anne Frank aspired to be taken seriously as a writer—and should be."
—*San Diego Union-Tribune*

"Lively and illuminating . . . an impressively far-reaching critical work, an elegant study both edifying and entertaining. In a book full of keen observations and fascinating disputes . . . Ms. Prose looks in all directions to find noteworthy material . . . This is a Grade A example of what a smart, precise and impassioned teacher can do." —Janet Maslin, *New York Times*

"One could reasonably ask if the world really needs yet another book about Anne Frank. Well, it probably needs this one . . . A sensitive, beautifully written and fascinating account of the myriad aspects of Anne Frank's life, death and diary."
—*Haaretz* (Israel)

"Substantially researched and wide-ranging . . . This probing and informed book introduces readers to a far more complex and accomplished young woman than the Anne we met in our adolescence . . . Prose examines Anne Frank as the writer she hoped to be, not as an iconic victim of the Holocaust."
—*Jewish Book World*

"Prose is commanding and illuminating . . . a definitive, deeply moving inquiry into the life of the young, imperiled artist. . . . An extraordinary testimony to the power of literature and compassion." —*Booklist* (starred review)

ABOUT THE AUTHOR

FRANCINE PROSE is the author of fifteen books of fiction, most recently the highly acclaimed *Goldengrove*. The novel *A Changed Man* won the Dayton Literary Peace Prize, and *Blue Angel* was a finalist for the National Book Award for Fiction. Her most recent work of nonfiction, *Reading Like a Writer*, was a *New York Times* bestseller. A former president of PEN American Center, she lives in New York City.

ANNE FRANK

The Book, The Life,

The Afterlife

FRANCINE PROSE

HARPER PERENNIAL

NEW YORK • LONDON • TORONTO • SYDNEY • NEW DELHI • AUCKLAND

HARPER ● PERENNIAL

FIRST HARPER PERENNIAL EDITION PUBLISHED 2010.

The Library of Congress has catalogued the hardcover edition as follows:

Prose, Francine
    Anne Frank : the book, the life, the afterlife / Francine Prose.
        p. cm.
    Includes bibliographical references.
    ISBN 978-0-06-143079-4
    1. Frank, Anne, 1929–1945. Achterhuis. 2. Frank, Anne, 1929–1945—Literary art.
3. Holocaust, Jewish (1939–1945)—Netherlands—Amsterdam—Personal narratives—
History and criticism. 4. Creative writing. 5. Frank, Anne, 1929–1945—Authorship.
I. Title.
    DS135.N6F73525 2009
    940.53'18092—dc22
                                                            2009017703

ISBN 978-0-06-143080-0 (pbk.)

10 11 12 13 14   OV/RRD   10 9 8 7 6 5 4 3 2 1

To Howie

CONTENTS

PART I

The Life

ONE

The Book, The Life, The Afterlife

I would call the subject of Anne Frank's *Diary* even more mysterious and fundamental than St. Augustine's, and describe it as: the conversion of a child into a person. . . . Why—I asked myself with astonishment when I first encountered the *Diary*, or the extracts *Commentary* published—has this process not been described before? universal as it is, and universally interesting? And the answer came. It is *not* universal, for most people do not grow up, in any degree that will correspond to Anne Frank's growing up; and it is not universally interesting, for nobody cares to recall his own, or can. It took, I believe, a special pressure forcing the child-adult conversion, and exceptional self-awareness and exceptional candour and exceptional powers of expression, to bring that strange or normal change into view.

—JOHN BERRYMAN, "The Development of Anne Frank"

She was a marvelous young writer. She was something for thirteen. It's like watching an accelerated film of a fetus sprouting a face, watching her mastering things.... Suddenly she's discovering reflection, suddenly there's portraiture, character sketches, suddenly there's long intricate eventful happenings so beautifully recounted it seems to have gone through a dozen drafts. And no poisonous notion of being *interesting* or *serious*. She just *is*.... The ardor in her, the spirit in her—always on the move, always starting things . . . she's like some impassioned little sister of Kafka's, his lost little daughter.

—PHILIP ROTH, *The Ghost Writer*

THE FIRST TIME I READ THE DIARY OF ANNE FRANK, I was younger than its author was when, at the age of thirteen, she began to write it. I can still picture myself sitting cross-legged on the floor of the bedroom in the house in which I grew up and reading until the daylight faded around me and I had to turn on the lamp. I lost track of my surroundings and felt as if I were entering the Amsterdam attic in which a Jewish girl and her family hid from the Nazis, and where, with the aid of their Dutch "helpers," they survived for two years and a month, until they were betrayed to the authorities, arrested, and deported. I was enthralled by Anne's vivid descriptions of her adored father, Otto; of her conflicts with her mother, Edith, and her sister, Margot; of her romance with Peter van Pels; and of her irritation with Hermann and Auguste van Pels and the dentist, Fritz Pfeffer, with whom the Franks shared the secret annex. I remember that when I finished the book, I went back to the first page and started again, and that I read and reread the diary until I was older than Anne Frank was when she died, at fifteen, in Bergen-Belsen.

In the summer of 2005, I read the diary once more. I had

just begun making notes for a novel that, I knew, would be narrated in the voice of a thirteen-year-old girl. Having written a book suggesting that writers seek guidance from a close and thoughtful reading of the classics, I thought I should follow my own advice, and it occurred to me that the greatest book ever written about a thirteen-year-old girl was Anne Frank's diary.

Like most of Anne Frank's readers, I had viewed her book as the innocent and spontaneous outpourings of a teenager. But now, rereading it as an adult, I quickly became convinced that I was in the presence of a consciously crafted work of literature. I understood, as I could not have as a child, how much art is required to give the impression of artlessness, how much control is necessary in order to seem natural, how almost nothing is more difficult for a writer than to find a narrative voice as fresh and unaffected as Anne Frank's. I appreciated, as I did not when I was a girl, her technical proficiency, the novelistic qualities of her diary, her ability to turn living people into characters, her observational powers, her eye for detail, her ear for dialogue and monologue, and the sense of pacing that guides her as she intersperses sections of reflection with dramatized scenes.

I kept pausing to marvel at the fact that one of the greatest books about the Nazi genocide should have been written by a girl between the ages of thirteen and fifteen—not a demographic we commonly associate with literary genius. How astonishing that a teenager could have written so intelligently and so movingly about a subject that continues to overwhelm the adult imagination. What makes it even more impressive is that this deceptively unassuming book focuses on a particular moment and on specific people, and at the same time speaks, in ways that seem timeless and universal, about adolescence and family life. It tells the truth about certain human beings' ineradicable desire to exterminate the largest possible number of other human beings, even as it celebrates the will to survive and the determination to maintain one's decency and dignity under the most dehumanizing circumstances.

Anne Frank thought of herself not merely as a girl who happened to be keeping a diary, but as a writer. According to Hanneli Goslar, a childhood friend, Anne's passion for writing began when she was still in school. "Anne would sit in class between lessons and she would shield her diary and she would write and write. Everybody would ask her, 'What are you writing?' And the answer always was, 'It's none of your business.'" In April 1944, four months before the attic in which the Franks found refuge was raided by the Nazis, Anne Frank recorded her wish to become a writer. "If I haven't any talent for writing books or newspaper articles, well, then I can always write for myself. . . . I want to go on living even after my death! And therefore I am grateful to God for giving me this gift, this possibility of developing myself and of writing, of expressing all that is in me!"

Much has been made of how differently we see Anne Frank after the so-called *Definitive Edition* of her diary, published in 1995, restored certain passages that Otto Frank had cut from the version that appeared in Holland in 1947 and in the United States in 1952. In fact, though the *Definitive Edition* is almost a third longer than the first published version of *The Diary of a Young Girl*, the sections that were reinstated—barbed comments about Edith Frank and the Van Pelses, and other entries revealing the extent of Anne's curiosity about sexuality and about her body—don't substantially change our perception of her.

On the other hand, there is a scene in Miep Gies's memoir, *Anne Frank Remembered*, that actually *does* alter our image of Anne. Along with the other helpers, the employees of Opekta, Otto Frank's spice and pectin business, Miep risked her life to keep eight Jews alive for two years and a month, an experience she describes in a book that sharpens and enhances our sense of what the hidden Jews and their Dutch rescuers endured. The scene begins when Miep accidentally interrupts Anne while she is at work on her diary.

> *I saw that Anne was writing intently, and hadn't heard me. I was quite close to her and was about to turn and go when she looked up, surprised, and saw me standing there. In our many encounters over the years, I'd seen Anne, like a chameleon, go from mood to mood, but always with friendliness. . . . But I saw a look on her face at this moment that I'd never seen before. It was a look of dark concentration, as if she had a throbbing headache. The look pierced me, and I was speechless. She was suddenly another person there writing at that table.*

The Anne whom Miep observed *was* another person: a writer, interrupted.

In his 1967 essay, "The Development of Anne Frank," John Berryman asked "whether Anne Frank has *had* any serious readers, for I find no indication in anything written about her that anyone has taken her with real seriousness." That is no longer completely true. In an incisive 1989 *New Yorker* essay, "Not Even a Nice Girl," Judith Thurman remarked on the skill with which Anne Frank constructed her narrative. A small number of critics and historians have called attention to Anne's precocious literary talent. In her introduction to the British edition of *The Tales from the House Behind*, a collection of Anne's fiction and her autobiographical compositions, the British author G. B. Stern wrote, "One thing is certain, that Anne was a writer in embryo." But is a "writer in embryo" the same as one who has emerged, at once newborn and mature?

The fact remains that Anne Frank has only rarely been given her due as a writer. With few exceptions, her diary has still never been taken seriously as literature, perhaps because it *is* a diary, or, more likely, because its author was a girl. Her book has been discussed as eyewitness testimony, as a war document, as a Holocaust narrative or not, as a book written during the time of war that is only tangentially about the war, and as a springboard for conversations about racism and intolerance. But it has hardly ever been viewed as a work of art.

Harold Bloom tells us why: "A child's diary, even when she was so natural a writer, rarely could sustain literary criticism. Since *this* diary is emblematic of hundreds of thousands of murdered children, criticism is irrelevant. I myself have no qualifications except as a literary critic. One cannot write about Anne Frank's *Diary* as if Shakespeare, or Philip Roth, is the subject."

The Dutch novelist Harry Mulisch attributed the diary's popularity to the fact that its young author died soon after writing it: "The work by this child is not simply *not* a work of art, but in a certain sense it is a work of art made by life itself: it is a found object. It was after all literally found among the debris on the floor after the eight characters departed. . . ." Writing in the *New Republic*, Robert Alter, a critic and biblical scholar, agreed: "I do not mean to sound impervious to the poignancy of the *Diary*. Still, many diaries of Jews who perished have been published that reflect a complexity of adult perspective and, in some instances, of a direct grappling with the barbarity of Nazism; and these are absent from Anne Frank's writing. . . . Anne may have been a bright and admirably introspective girl, but there is not much in her diary that is emotionally demanding, and her reflections on the world have the quality of banality that one would expect from a 14-year-old. What makes the *Diary* moving is the shadow cast back over it by the notice of the death at the end. Try to imagine (as Philip Roth did, for other reasons, in *The Ghost Writer*) an Anne Frank who survived Bergen-Belsen, and, let us say, settled in Cleveland, became a journalist, married and had two children. Would anyone care about her wartime diary except as an account of the material circumstances of hiding out from the Nazis in Amsterdam?"

At once admiring of Anne's gifts and troubled by a sense of how they have been underestimated, I began to think it might be interesting and perhaps useful for students newly introduced to Anne's diary and for readers who have grown accustomed to seeing it in a certain light to consider her work from a more literary perspective. What aspects of the book have helped to ensure its long and influential afterlife? Why has Anne Frank become such an iconic figure for so many readers, in so many countries? What is it about her voice that continues to engage and move her audience? How have the various interpretations and versions of her diary—the Broadway play, the Hollywood film, the schoolroom lessons, the newspaper articles that keep her in the public eye—influenced our idea of who she was and what she wrote?

The book I imagined would address those questions, mostly through a close reading of the diary. Such a book would explore the ways in which Anne's diary found an enduring place in the culture and consciousness of the world. I would argue for Anne Frank's *talent as a writer*. Regardless of her age and her gender, she managed to create something that transcended what she herself called "the unbosomings of a thirteen-year-old" and that should be awarded its place among the great memoirs and spiritual confessions, as well as among the most significant records of the era in which she lived.

That was the simple little book I envisioned. But little about the diary would turn out to be so simple.

I HAD always believed Anne Frank's diary to be a printed version (lightly edited by her father) of the book with the checked cloth cover that she received on her thirteenth birthday in June 1942, and that she began to write in shortly before she and her family went into hiding. That was what I had assumed, especially after I, like the rest of Anne's early readers, had been reassured by the brief epilogue to early editions of her book, in which we were informed that "apart from a very few passages, which are of little interest to the reader, the original text has been printed."

I knew there had been controversies about the missing pages Otto Frank had omitted in the process of shaping the diary. More recently, I recalled, more withheld pages had surfaced, passages in which Anne speculated about the disappointments in her parents' marriage. But I had thought that these questions had been answered, and most of the cuts restored, in the 1995 publication of the *Definitive Edition,* edited by Mirjam Pressler.

In fact, as I soon learned, Anne had filled the famous checked diary by the end of 1942; the entries in the red, gray, and tan cloth-covered book span the period from June 12, 1942, until December 5 of that year. Then a year—that is, a year of original, unrevised diary entries—is missing. The diary resumes in an exercise book with a black cover, which the Dutch helpers brought her. Begun on December 22, 1943, this continuation of the diary runs until April 17, 1944. A third exercise book begins on April 17, 1944; the final entry was written three days before its writer's arrest on August 4.

Starting in the spring of 1944, Anne went back and rewrote her diary from the beginning. These revisions would cover 324 loose sheets of colored paper and fill in the one-year gap between the checked diary and the first black exercise book. She continued to update the diary even as she rewrote the earlier pages. Anne had wanted her book to be noticed, to be read, and she spent her last months of relative freedom desperately attempting to make sure that her wish might some day be granted.

ON March 29, 1944, the residents of the secret annex gathered around their contraband radio to listen to a broadcast of Dutch news from London. In the course of the program, Gerrit Bolkestein, the minister of education, art, and science in the exiled Dutch government, called for the establishment of a national archive to house the "ordinary documents"—diaries, letters, sermons, and so forth—written by Dutch citizens during the war. Such papers, the minister said, would help future generations understand what the people of Holland had suffered and overcome.

As they listened, the eight Jews in the annex focused on the young diarist in their midst. "Of course they all made a rush at my diary immediately." *Of course.* Anne's diary was a fact of communal life, like the potatoes they ate, the bathing arrangements they worked out, the alarming break-ins downstairs, and it inspired curiosity and consternation in the people she was writing about. As early as September 1942, Anne describes snapping her little book shut when Mrs. Van Pels comes into the room and asks to see the diary, in which Anne has just been writing about her, unflatteringly. A month later, during a moment of closeness—Margot and Anne have gotten into the same bed—Margot asks if she can read Anne's diary, and Anne replies, "Yes—at least bits of it."

For most of her stay in the annex, Anne's diary had been her friend and her consolation. She wrote it for companionship, for the pleasure of writing, for a way to help fill the long hours during which she and the others were required to keep silent and nearly motionless while business was being transacted in the Opekta office downstairs. She wrote to help make sense of herself and the people around her. As Philip Roth notes in *The Ghost Writer,* the diary "kept her company and it kept her sane."

But now, at this hopeful juncture, when it had begun to

seem that the war might end and that people might want to read about the lives of its victims and its survivors, the attic residents agreed that Anne's diary was exactly the sort of thing the exiled Dutch minister meant. Anne took his speech as a personal directive. By the morning after the broadcast, she was envisioning a bright career for her book, a future more glamorous than what Minister Bolkestein proposed: posterity in the archive that would become the Netherlands Institute for War Documentation.

"Just imagine how interesting it would be if I were to publish a romance of the 'Secret Annex.' The title alone would be enough to make people think it was a detective story. But, seriously, it would be quite funny ten years after the war if we Jews were to tell how we lived and what we ate and talked about here." The title Anne had in mind, *Het Achterhuis*—literally, "the house behind" or "the annex"—refers to the fact that the rooms in which she and her family hid were above Otto Frank's former workplace and concealed from the street by the buildings around it. Many old Dutch houses had annexes of this sort, a maze of extra rooms added onto the back of the house, meant to extend the cramped space dictated by the structure's narrow facade.

A few days later, Anne lay on the floor and sobbed until the idea of herself as a writer lifted her out of despair. "I must work, so as not to be a fool, to get on to become a journalist, because that's what I want! I know that I can write, a couple of my stories are good, my descriptions of the 'Secret Annex' are humorous, there's a lot in my diary that speaks, but—whether I have real talent remains to be seen. . . ."

On April 14, she had serious misgivings about her abilities. Even so, she was imagining the Dutch ministers as her potential audience, and her critics: "Everything here is so mixed up, nothing's connected any more, and sometimes I very much doubt whether anyone in the future will be interested in all my tosh. 'The unbosomings of an ugly duckling' will be the title of all this nonsense; my diary really won't be much use to Messrs. Bolkestein or Gerbrandy."

In May she again wrote that she wished to become a journalist and a famous author—only now she had a sense of the book that might make her reputation. "Whether these leanings towards greatness (insanity!) will ever materialize remains to be seen, but I certainly have the subjects in my mind. In any case, I want to publish a book called *Het Achterhuis* after the war. Whether I shall succeed or not, I cannot say, but my diary will be a great help."

The most important result of this new sense of vocation was that Anne began to refine and polish her diary into a form that she hoped might someday appear as *Het Achterhuis*. On May 20, she wrote, in a passage her father deleted, "At long last after a great deal of reflection I have started my 'Achterhuis,' in my head it is as good as finished, although it won't go as quickly as that really, if it ever comes off at all."

In *The Ghost Writer*, Roth's hero, Nathan Zuckerman, remarks that the diary's dramatic scenes seemed to have gone through a dozen drafts. The truth is that many of them *did* go through at least two.

Returning to the earliest pages, Anne cut, clarified, expanded her original entries, and added new ones which in some cases she predated, sometimes by years. Thus the book is not, strictly speaking, what we think of as a diary—a journal in which events are recorded as they occur, day by day—but rather a memoir in the form of diary entries. The translator of the *Definitive Edition*, Mirjam Pressler, has written one of the few books that acknowledges the importance of Anne's revisions. Published in English as *Anne Frank: A Hidden Life*, and, oddly, targeted at a young-adult readership, Pressler's book mixes biographical information, a meditation about Anne and the others in the annex, and illuminating comparisons between the original diary and the version Anne rewrote. *"The Diary of a Young Girl* is not a diary kept in chronological order from beginning to end as one might expect. The main part of the book consists of the second version of Anne's original diary, revised with additions by Anne herself, with some stories from the account book in which she also wrote."

Judith Thurman got it right, as few have, when she questioned even calling the book, as Anne's American publishers did, *The Diary of a Young Girl*. "That ingenuous title corresponds to what is in fact an epistolary autobiography of exceptional caliber. It takes the full measure of a complex, evolving character. It has the shape and drama of literature. It was scrupulously revised by its author, who intended it to be read. It is certainly not a piece of 'found art,' as one Dutch critic has suggested."

One can understand Doubleday's belief that *The Diary of a Young Girl* was a catchier title than *The House Behind*. Though Anne Frank imagined *Het Achterhuis* as a novel in the form of a journal, it has come down to us as a diary. In *The Ghost Writer*, Philip Roth—who, as a fellow novelist, would be naturally sensitive to a writer's prerogative to call her book what she wants—refers to Anne's book only as *Het Achterhuis*, and to the Broadway play by *its* name, *The Diary of Anne Frank*.

DESPITE Anne's initial misgivings, the revision of *Het Achterhuis* went very quickly. Correlating the penmanship of the loose sheets against that of the notebooks, the forensic handwriting analysts later employed by the Netherlands Institute for War Documentation concluded that "if we take May 20, 1944, as the starting date (on the basis of the comment in part 3) and August 1, 1944, as the date of the last entry, then the average daily entry would run to from 4 to 5 pages a day. These must have been written in addition to the entries in the diary, part 3. . . . It appears that the writer worked more intensely on the loose sheets, particularly in the period between July 15 and August 1, 1944. During that period, 162 pages were completed, or about 11 pages a day."

Working at this astonishing rate, Anne rewrote her early draft in the weeks before her arrest, making major and minor changes. Like any memoirist fearing hurt feelings, or accusations of misrepresentation, she made a list of pseudonyms for the Jews and their helpers. The Frank family would become the Robins, the Van Pelses would be called the Van Daans, while the dentist, Fritz Pfeffer, would appear in the book as Albert Dussel. Perhaps for fluency, she continued to use the real names when she wrote her second draft.

"I am the best and sharpest critic of my own work. I know myself what is and what is not well written. Anyone who doesn't write doesn't know how wonderful it is." By the time she made her final entry, on August 1, 1944, she had revised the passages that preceded the March radio broadcast and kept the diary up to date in an unrevised first draft.

After the war, when Otto Frank read over his daughter's work and became convinced that she'd meant it to be published, he prepared a version of the book that combined passages from Anne's first draft and from her revisions, in some cases using earlier versions of passages that she had subsequently revised. All in all, Otto Frank did an admirable job of editing—omitting needless details, choosing between alternate versions of events, preserving the essence of the diary, and intuiting what would make the book more appealing to readers. In many cases, that meant reversing Anne's decisions about what she wanted omitted—for example, the intensely emotional entries from the start of her romance with Peter van Pels, with whom she had become disenchanted during the time she was rewriting her diary.

The cooling of the love affair and Anne's focus on the revisions may not be entirely unrelated. Once she had stopped thinking semiobsessively about the boy upstairs, Anne had more time and energy to devote to her writing. She would not have been the first artist to discover that the end of a romance can inspire a return to work with new energy and sharpened concentration.

In 1986, the Netherlands Institute for War Documentation published *The Critical Edition of the Diary of Anne Frank*, a huge volume, over eight hundred pages long, that includes all the extant drafts of Anne's diary; the English edition would appear three years later. Missing from the book were the five pages that Otto Frank and the Frank family chose to leave out, pages that subsequently appeared in the more recent *Revised Critical Edition*, which was published in Dutch in 2001 and in English in 2003. Both the earlier and later editions contain an account of the methods and conclusions of the forensic experts employed by the institute, who proved that the diaries, except for a few minor editorial corrections, were written entirely by Anne Frank. Their meticulous research demonstrated how the evolution of Anne's handwriting over the course of the two years in hiding took the exact trajectory that the penmanship of a child—the same child—would be expected to follow between the ages of thirteen and fifteen.

In *The Critical Edition*, the original draft of Anne's diary is referred to as the "a" version. The revisions that she made on the loose sheets constitute the "b" version. And the book that her father produced by combining those first two drafts is reprinted as the "c" version. All three drafts are printed in parallel bands, so that it is possible—painstaking, time-consuming, and at times maddening, but possible—to read all three versions and to determine what Anne originally wrote, what she rewrote, what she intended to appear in *Het Achterhuis*, and at what points her father respected or reversed her decisions. Judith Thurman observed, "What a comparison of the texts does reveal is both how spontaneously the diarist composed her prose and how finely she then tuned it. In order to make such a comparison, however, one needs a certain amount of motivation. The editors' instructions on how to read *The Critical Edition* are more arcane, and harder to follow, than those for a build-it-yourself hang glider."

What makes the task of comparing Anne's original draft with her revisions and with her father's compilation even more challenging are all the unanswered and unanswerable questions. When Anne said that she had begun writing *Het Achterhuis*, did she mean that she had *just* begun? Long gaps in each version must be filled in by consulting the others. Even in the cloth diary, pages are misnumbered and dated out of order. If Anne omitted something from her second draft, did that mean that she intended to excise it completely, or that she felt the first version was sufficient? And finally, for those of us who don't read Dutch, there is the problem of knowing how much we are missing by reading the work in translation. According to David Barnouw, one of the editors of *The Critical Edition*, only readers of Dutch can appreciate how much Anne's style changed over those two years. In the published version, explains Barnouw, Anne's incorrect word choices and other youthful mistakes were rectified, and the rougher passages were smoothed out. "Otherwise, it would have seemed that the editor made a mistake."

One of the most clear-sighted experts on the diary is Laureen Nussbaum, a childhood acquaintance of the Frank family (after the war, Otto would be the best man at her wedding) who went on to become a professor at the University of Oregon. She was the first to note that the revised, or "b," draft—Anne's own version of the text—has never been published as a stand-alone volume. The 1995 *Definitive Edition*, commented Nussbaum, only further muddied the waters since many of the cuts it restored (Anne's reflections on her sexuality and outbursts of rage at her mother) were sections that Anne herself had removed from the book she hoped to publish.

As soon as I had taken the time to understand what *The Critical Edition* contained, and the implications of the alternative versions, I realized that to write about Anne Frank as an artist would be more involved than a straightforward close reading of *The Diary of a Young Girl*. Suddenly, it was as if Anne had written two books—at least two books—that needed to be considered.

The first was the diary that has had, and continues to have, such a powerful effect on readers, the book that has been adapted for Broadway and Hollywood, and that is still taught in classrooms everywhere. That is the so-called "c" version. Any discussion of Anne's influence, of the intensity with which her diary has been taken to heart by its fans, and of the ways in which her "message" has been interpreted requires us to look at the "c" version (regardless of what we might think of the successive drafts, or of Otto Frank's editing) as if it *were* the only version of the diary. Which, in effect, it is—except for the few readers, among millions, who have professional or private reasons for wanting to study the published diary alongside the alternate drafts.

At the same time, it seemed unfair to Anne Frank as a writer to ignore what the variant drafts provide: evidence of her creative process, of her gifts for revision, of her first and second thoughts about how she wanted to portray herself and those around her. What was added to, and lost from, the book during those final months as Anne feverishly rewrote—on the loose colored sheets—the observations, reflections, and self-representation of an earlier self?

But that was only part of the complexities involved. When I began to consider writing about the diary, I had only a vague notion of the controversies it had inspired. I knew that Anne's work and her symbolic significance have incited battles extending far beyond the book itself. I had heard that the diary's journey from the printed page to the stage and screen was a rocky one, but I'd had no idea that it involved lawsuits, betrayals and alliances, accusations of plagiarism and breach of contract, and obsessive paranoia concerning Zionist or Stalinist plots. Few other writers have given rise to such intense emotion, such fierce possessiveness, so many arguments about who is entitled to speak in her name, and about what her book does, and doesn't, represent. Few have had such an effect on the world, and inspired the sort of devotion that more often surrounds the figure of a religious leader, or a saint.

If one hallmark of a masterpiece is the burrlike tenacity with which it sticks in our memory, Anne Frank's journal claims that status as a consequence of the indelible impression that its "plot" and "characters" leave on its readers. Decades after the diary's publication, the entrance to the secret annex remains the door through which new readers, many of them young, will first enter the historical moment in which it was written. When the book is taught in classrooms everywhere, all sorts of lessons—frequently improving, occasionally peculiar, and often quite unlike anything Anne Frank could have intended—are extracted from its pages. Anne's diary is one of the texts most frequently read and studied by incarcerated men and women in prisons throughout the United States.

The book has been translated into dozens of languages; tens of millions of copies are in print. The extent to which the figure of Anne Frank has permeated world culture can perhaps be seen in the fact that, in Japan (where the book was an enormous success, selling 116,000 copies in its first five months in print), to have one's "Anne Frank day" became a euphemism for menstruation, a subject Anne mentioned in her journal. A variety of rose named after Anne Frank now grows all over Japan.

A further measure of the book's currency is the excitement generated by each new revelation about Anne's life or about her diary. On September 10, 1998, the *New York Times* ran a two-thousand-word essay, beginning on the front page, headlined "Five Precious Pages Renew Wrangling over Anne Frank" and subtitled: "A long-withheld page from Anne Frank's diary reveals difficulties with her mother: 'I am unable to talk with her. I cannot look lovingly into those cold eyes. I cannot, never!'" It's hard to think of another literary text—a lost Shakespeare sonnet? a previously unknown verse of the Bible?—whose discovery would have received such prominent coverage, especially if the passage concerned a young girl's view of her parents' marriage.

Yet another major news story broke in 2005 when a cache of letters was discovered at New York's YIVO Institute, correspondence documenting Otto Frank's desperate attempts to find asylum for his family in the United States or Cuba. These letters inspired a Long Island congressman to campaign—in vain—to have Anne Frank granted honorary U.S. citizenship as partial atonement for our government's refusal to save the Franks.

A range of films and plays have attempted to tell Anne's story, with varying degrees of success. Jon Blair's *Anne Frank Remembered* won an Academy Award for Documentary Feature in 1996. Films and docudramas have included "re-creations" in which actors played the Franks and their neighbors, and one made-for-television film, *Who Betrayed Anne Frank?*, frames the story as a detective procedural with the sort of ominous sound track we associate with shows about the riddle of the Mayans' disappearance. At the 2007 New York Fringe Theater festival, *Days and Nights: page 121, lines 11 and 12* featured actors recognizable as the characters in Anne's diary, but who—in Marc Stuart Weitz's play—passed their time in the attic reciting Chekhov's *The Seagull*. The 2003 hip-hop film, *Anne B. Real*, centers on a female rapper who finds inspiration in Anne Frank's story, while the popular book *The Freedom Writers Diary* and the subsequent film, *Freedom Writers*, describes how an inner-city classroom was energized by a journal kept during a war that few of the students had known much about. Anne's story has even been made into a Japanese anime cartoon, *Anne no Nikki*.

In 1998, the indie band Neutral Milk Hotel released *In the Aeroplane over the Sea*, an album of songs partly inspired by Anne Frank's life and death. Ten years later, a musical adapted from the diary—*The Diary of Anne Frank: A Song to Life*—opened in Madrid. A puppet show of the diary has appeared to sell-out crowds in Atlanta, while an episode of *60 Minutes* reported that North Korean schoolchildren were being assigned to read Anne's journal with instructions to think of George W. Bush as Hitler and of the Americans as the Nazis who wished to exterminate the North Koreans.

Books of nonfiction and fiction have expanded upon, and been inspired by, what Anne confided to her journal. Periodically, the publishing industry discovers the war diary of some hapless young person and promotes its author as the Anne Frank of Serbia, or Poland, or Vietnam, or the latest place where children are the victims of their elders. Philip Roth's 1979 novel, *The Ghost Writer*, includes a sustained meditation about Anne and her diary occasioned by Nathan Zuckerman's fantasy that the beautiful mistress of his literary idol is Anne Frank, who has not only survived the camps but has come to America, where she is living under a pseudonym and has landed a job archiving manuscripts for her lover. Roth's Anne Frank character, Amy Bellette, reappears, older and infirm, in his 2007 novel, *Exit Ghost.*

That same year, newspapers around the world reported that, weakened by age and disease, the chestnut tree outside the secret annex was in danger of being cut down. Emotions ran high during the debate about whether the leafy messenger that had brought Anne news about the changing seasons could be saved. As I write this, the valiant old tree struggles on in the courtyard of the former warehouse where the Frank family hid, and plans are being made to import and plant ten saplings from the tree in the United States.

TWO

~

The Life

ONE PROBLEM CONFRONTING EVERY WRITER OF FICTION
or nonfiction is the question of background. How much must a
reader know in order to make sense of what the author is trying
to convey? In a diary entry dated June 20, 1942, but written
almost two years afterward, Anne acknowledges the necessity
of giving Kitty, her invented confidante, enough information to
enable her to follow the narrative. "I don't want to set down a
series of bald facts in a diary like most people do . . . but no one
will grasp what I'm talking about if I begin my letters to Kitty
just out of the blue, so I'll start by sketching in brief the story
of my life."

We'll return to this entry later, but, in passing, let's note the
phrase: "no one will grasp what I'm talking about if I begin my
letters to Kitty just out of the blue." Not only does it suggest
that this is something other than a girl confiding in her diary,
but it contradicts what Anne says in the same entry: "I don't
intend to show this cardboard-covered notebook, bearing the

proud name of 'diary,' to anyone, unless I find a real friend, boy or girl, probably no one cares."

The sketch that ensues, presumably intended for that "real friend," and in truth for a wider audience, could hardly be more economical or concise. Anne begins by explaining that her father was thirty-six when he married her mother, who was twenty-five, that Anne's sister, Margot, was born in 1926 in Frankfurt am Main, and that Anne herself—Annelies Marie Frank—was born three years later, on June 12, 1929.

In May 1944, Anne asks Kitty if she has ever really told her anything about her family and proceeds to flesh out her earlier outline. She explains that her father was born in Frankfurt am Main, where her grandfather Michael Frank owned a bank. As a boy, Otto attended dances, and there were parties every week. Surrounded by beautiful girls, he enjoyed waltzing and lavish dinners. After her grandfather's death, much of the money was lost; war and inflation took what was left. Her mother wasn't quite so rich, Anne informs Kitty, but still there was plenty of money, and Edith often delighted her daughters with stories about engagement parties attended by 250 guests.

Anne's uncharacteristic longing for her parents' lost wealth and their privileged childhoods has been precipitated by the deterioration of living conditions in the annex. Since the arrest of Miep's trusted black-market coupon dealer, the hidden Jews have either starved or been forced to eat spoiled food. In case their diet isn't demoralizing enough, Miep—meaning well, as always—has tried to cheer them up with a story about an engagement party she attended. At the celebration, her hosts served vegetable soup with meatballs, cheese, rolls, roast beef, cakes, and wine; this enviable menu inspired, in the hungry young writer, an uncharacteristic joke at the expense of her beloved helpers ("Miep had ten drinks and smoked 3 cigarettes; can that be the woman who calls herself a teetotaler? If Miep had all those, I wonder how many her spouse managed to knock back?") that Otto cut from the published diary.

In any case, Miep's description led Anne to compare the delights outside the annex with the privations inside it, and the present with the past. "Miep made our mouths water telling us about the food they had. . . . We, who get nothing but two spoonfuls of porridge for our breakfast and whose tummies were so empty that they were positively rattling, we, who get nothing but half-cooked spinach (to preserve the vitamins!) and rotten potatoes day after day, we, who get nothing but lettuce, cooked or raw, spinach and yet again spinach in our hollow stomachs. Perhaps we may yet grow to be as strong as Popeye, although I don't see much sign of it at present!

"If Miep had taken us to the party we shouldn't have left any rolls for the other guests. If we had been at the party we should undoubtedly have snatched up the whole lot and left not even the furniture in place. . . . And these are the granddaughters of a millionaire. The world is a queer place!"

THOUGH not quite the millionaire his granddaughter imagined, Michael Frank was the founder of the Bank Michael Frank, based in Frankfurt, where Otto grew up in a close-knit, assimilated German-Jewish community, surrounded by art and good furniture. Servants. *Parties every week.*

After one semester at Heidelberg, Otto left the university and traveled to New York with a school friend, Nathan Straus, whose family owned Macy's department store. Otto worked at the store until, in 1909, he was called back to Germany to deal with the family finances in the aftermath of his father's sudden death. Like his brothers Herbert and Robert, Otto served in the German army during World War I. As part of a range-finding unit, Otto fought with an infantry corps composed largely of surveyors and mathematicians. By the time the war ended, he had been promoted to lieutenant, and in 1925 he married Edith Hollander, whose father ran a successful business dealing in scrap iron.

Otto spent his early adulthood attempting to save the family bank as it gradually went under, weakened by political and personal crises: the war, hyperinflation, the stock market crash, the Great Depression, a scandal in which Otto's brother Herbert was accused of illegal dealings in foreign securities, and the end of Weimar democracy.

Though he has been charged with an ostrichlike refusal to understand the implications of the rise of National Socialism and to foresee the threat it would pose to his livelihood and his family, in fact Otto had a talent for apprising his situation and for operating under stress. Many years later, the producer and playwrights who brought Anne's diary to Broadway remarked that Otto was not only the loving, grieving father of a murdered girl, but a gifted businessman who grasped the practical and financial ramifications of her diary's success.

Having opened and then liquidated a branch of the Michael Frank bank in Amsterdam, in the 1920s, Otto knew and liked the Dutch capital. He had made contacts there who would prove useful when, in 1933, Hitler was appointed chancellor and the Nazis' increasingly vicious anti-Jewish laws convinced him that the wisest option was to leave Germany and move his wife and daughters to the deceptive safety of Holland.

Anne concludes the passage about the engagement party that Miep describes and about the Frank family's fall from grace: "Daddy was therefore extremely well brought up and he laughed very much yesterday, when, for the first time in his fifty-five years, he scraped out the frying pan at the table."

THE ENTRY dated June 20, 1942, continues: ". . . *as we are Jewish, we emigrated to Holland in 1933 . . .*" Like any skilled writer wisely determined to omit unnecessary details, Anne brings us directly to the emigration that became urgent *as we are Jewish.*

In 1933, around the time that Nazi storm troopers initiated a boycott of German-Jewish businesses, the Franks bid goodbye to Frankfurt. Otto shut down the Michael Frank bank, left Edith and the girls with his mother-in-law in Aachen, and went ahead to Amsterdam. There, with the help of his brother-in-law, Erich Elias, who had emigrated to Basel and was employed by the Swiss branch of a German company selling jelling agents for jams and preserves, Otto established a branch of the Opekta pectin supplier, with a limited marketing range restricted to private customers. A few years later, Otto formed an additional company, Pectacon, trading in seasonings and spices. He had been raised to run a bank, but business was business, and he could adapt. He could protect and support his family, which, for Otto, was always the highest priority.

A short time later, Edith—who had made several trips to help Otto find new quarters—rejoined her husband. In December, Margot was reunited with her parents in Amsterdam. In February 1934, the Franks decided that Anne should appear as the surprise birthday present for her older sister; the family story was that little Anne was plunked on the table as a gift. So she joined the rest of her family in their home at 37 Merwedeplein, in a newly developed South Amsterdam district, the River Quarter, which had become a magnet for German-Jewish refugees.

The neighborhood, and the Frank home, became the center of a community. The parents of Anne's playmates visited on weekends and holidays. On Purim, in either 1938 or 1939, as reports of the Nazi anti-Jewish violence in Germany were growing more disturbing, the father of one of Anne's friends startled and amused the other parents by dressing up as Hitler and standing at full attention when they came out to see who had rung the doorbell.

Already a reader, Anne was enrolled in the progressive Montessori kindergarten, a short walk from the Franks' new apartment. Soon, far more rapidly than her mother, she learned to function in a new language.

A demanding and often sickly baby, Anne grew into a challenging child—mercurial, moody, humorous, alternately outgoing and shy. A natural performer, she liked to pop her elbow out of its socket to get her friends' attention. She was bossy, theatrical, and outspoken. She was only four when she and her beloved grandmother Oma Hollander boarded a crowded Aachen streetcar, and Anne demanded, "Won't someone offer a seat to this old lady?"

In Amsterdam, she grew close to Hanneli Goslar, the "Lies" about whom Anne would later have the waking nightmare she describes in the diary. ("I saw her in front of me, clothed in rags, her face thin and worn.") A German refugee who had arrived in Holland around the same time as Anne, Hanneli met Anne in a grocery store; their mothers were glad to find someone with whom they could speak German. The Franks called on Hanneli Goslar's parents every Friday evening, and the two families celebrated Passover together. Eventually, Hanneli's mother, Ruth, would say about Anne, "God knows everything, but Anne knows everything better."

At the Montessori kindergarten, the prevailing theory was that adults should encourage children to flourish and grow and have a voice in deciding what they wished to do—and what sort of person they wanted to become. Otto and Edith Frank agreed; later, in the annex, the Van Pelses would often criticize the Franks for their "modern" ideas about childrearing.

By nature more lenient than his wife, Otto was, perhaps consequently, more popular, not only with his daughters but also with the girls' friends. Good looking, tall, patient, and courtly, Otto was the kind of father who taught the neighborhood children how to ride their bikes. How one pities the conventional, anxious Edith Frank, not confident, stiff, and far outmatched by her firecracker of a daughter, whom her husband adored.

The day after she and Anne met in the grocery, Hanneli began at the Montessori kindergarten, where, not knowing the language or any of the other children, she was hugely relieved to see Anne, from the back, playing music with bells. Anne turned, saw Hanneli, ran over to her, and threw her arms around her. "From then on we were friends." Anne's friendships, like those of many girls her age, had the intensity of love affairs, with all the concomitant jealousies, quarrels, separations, and reconciliations. Her high spirits and affectionate, impulsive generosity put her at the center of a tight clique that included Hanneli Goslar and Susanne Lederman. Their exclusive little trio was known, in their neighborhood, as Anne, Hanne, and Sanne.

Eva Geiringer-Schloss, Anne's near neighbor on Merwedeplein, arrived from Vienna, via Brussels, in 1940. After the war, her mother, Fritzi, would marry the widowed Otto Frank. In her memoir, *Eva's Story*, Anne's former classmate describes the "inseparable" Anne-Hanne-Sanne troika as being more sophisticated, more like teenagers, than the other girls, whom the chosen three—giggling about boys and fashion and film magazines—viewed with barely concealed disdain. They were famously boy crazy, especially Anne. One friend remembered Anne assuming that every boy wanted to be her boyfriend. Hanneli Goslar remarked that Anne was "always fussing" with her long hair. "Her hair kept her busy all the time."

Eva's memories of the enviably stylish Anne Frank include this revealing story:

Once, when Mutti had taken me to the local dressmaker to have a coat altered, we were sitting waiting our turn and heard the dressmaker talking to her customer inside the fitting room. The customer was very determined to have things just right.

"It would look better with larger shoulder pads," we could hear her saying in an authoritative tone of voice, "and the hemline should be just a little higher, don't you think?"

We then heard the dressmaker agreeing with her and I sat there wishing I was allowed to choose exactly what I wanted to wear. I was flabbergasted when the curtains were drawn back and there was Anne, all alone, making decisions about her own dress. It was peach-coloured with a green trim.

She smiled at me. "Do you like it?" she said, twirling around.

Interviewed by Ernst Schnabel, a novelist and dramatist who served in the German navy during World War II and who wrote the 1958 book *Anne Frank: A Portrait in Courage*, the mother of Anne's friend Jopie van der Waal (Schnabel employed the pseudonym used in Anne's diary for Jacqueline van Maarsen) also remembered making dresses for Anne. But what she mostly recalled is Anne's forceful personality, her desire to be a writer, and her precocious sense of self. The phrase, "She knew who she was," recurs, like a refrain, throughout the conversation, during which Mme. Van der Waal described the ceremony and the theater with which Anne arrived to spend the weekend:

"When Anne came to stay with us, she always brought a suitcase. A suitcase, mind you, when it wasn't a stone's throw between us. The suitcase was empty of course, but Anne insisted on it, because only with the suitcase did she feel as if she were really traveling."

A FLICKER of a home movie. June 22, 1941. The whole thing lasts ten seconds.

The bicycles slipping by provide the only indication that we are in Holland. The brick Merwedeplein apartment block looks more like married students' housing on an American state university campus than the quaint center-city canal houses we associate with Amsterdam.

The camera waits outside a door, peering up a stairwell. In search of something to focus on, it pans up the side of a building. In the open windows are neighborhood residents, girls and young women, their elbows propped on the sills, waiting. The women at the windows alter the look of the street, so the scene begins to look more like a village in southern Europe.

The newlywed couple appears, arm in arm, the groom in a top hat, cane, and formal wear, the bride in a flattering pale suit, a jaunty white fedora, and gloves; she carries a bouquet. They walk down the stairs and pause like movie stars obliging the paparazzi. Passersby lean against their bicycles, staring.

Suddenly, the camera zooms toward the sky and finds Anne Frank, watching from her window. She turns and speaks to someone inside the apartment. She looks back at the couple, then away. The camera appears to lose interest. It glances at a few more spectators, then returns to the Amsterdam street.

On the Web site for the United States Holocaust Memorial Museum, you can watch those few seconds of Anne on film, in blurred and grainy close-up. Anne's body language is quick, electric. A breeze, or maybe the motion of her body, lifts her hair as she turns, and her eyes smudge into dark ovals as she gazes down at the bridal couple.

As familiar as we are with images of Anne Frank, as inured as we may think we are to the sight of her beautiful face, the film pierces whatever armor we imagine we have developed. It is always shockingly short and always the same, and yet you are never entirely sure what you have, or haven't, seen. It's less like watching a film clip than like having one of those dreams in which you see a long-lost loved one or friend. In the dream, the person isn't really dead. You must have been mistaken. You wake up, and it takes a few moments to understand why the dream was so cruelly deceptive.

The film was shot by a friend of the groom, who still had the footage when Ernst Schnabel interviewed him for *Anne Frank: A Portrait in Courage*. The groom, identified only as Dr. K., showed it on a screen with a home-movie projector. Dr. K. explained that he didn't know Anne Frank, and that his wife, the bride in the film, "knew her only from Anne's girlhood days on Merwedeplein, simply as one knows the children of neighbors, from seeing them on the street and greeting them in the early morning. The friend who had filmed their wedding also did not know Anne, and the doctor guesses that there was a small strip of film left on the reel, not enough to do anything with, and so his friend had simply taken a shot into the blue. He had certainly never imagined that out of the blue he would catch in his lens ten seconds of history."

FROM the June 20, 1942, entry:

> *The rest of our family who were left in Germany felt the full impact of Hitler's anti-Jewish laws, so life was filled with anxiety. In 1938 after the pogroms, my two uncles (my mother's brothers) escaped to North America. My old grandmother came to us, she was then seventy-three. After May 1940 the good times rapidly fled: first the war, then the capitulation, followed by the German invasion, which is when the sufferings of us Jews really began.*

Otto Frank was smart enough to leave Germany a full five years before the orgy of broken glass, intimidation, and gang violence that would become known as Kristallnacht. And he reached Holland before the Dutch had had time to become alarmed by the growing number of Jewish refugees streaming across the German border.

By 1939, when Edith's mother finally left Germany and came to live with the Franks, the influx of Jews was so heavy that the Dutch government decided to build a resettlement station in the center of the country, a site to which the queen, Wilhelmina, objected because it was too near her estate. The proposed camp, which was to be financed by Jewish organizations, was relocated to Westerbork, in the raw, cold, sandy, fly-infested northeast. Later, as simply as reversing the hinges on a door, the pestilential camp was turned into a detention center for Jews being shipped out of the country by the Nazis. Tens of thousands, including the Franks and Van Pelses, were deported from Westerbork to their deaths in the east.

The Dutch remained so convinced that their neutrality would be respected that when the German invasion began, on May 10, 1940, even savvy Dutch journalists failed to pay attention to the evidence crashing around their ears. "All the correspondents report strange noises that have been audible along the border since nightfall. The heavy droning of motors, explosions, and other noises harder to identify. Also angry barking, apparently from startled farm dogs, and the lowing of restless cattle."

The battle lasted five days. Near the end, the Nazis threatened to bomb Rotterdam if the Dutch refused to surrender, and when negotiations stalled, the Nazis carried out their threat, killing nine hundred people. The Dutch rapidly capitulated. Queen Wilhelmina attempted to join the resistance in the south, but was persuaded to escape to Great Britain and form a government in exile. After trying to flee by boat, more than a hundred and fifty Jews committed suicide on May 15 and 16, when the Germans marched into Amsterdam. That was when, as Anne wrote four years later, in the phrase that the fifteen-year-old found to express the innocence of her younger self, "the good times rapidly fled."

In fact, the good times lasted slightly longer in the Netherlands than they did in other Nazi-occupied countries, in part because the invaders wished to preserve good relations with the Dutch, fellow Aryans whom they hoped might welcome the chance to join an ethnically pure greater Germany. But despite the gradual pace at which the Nazis implemented anti-Jewish regulations in Holland, their intentions soon became clear.

By that fall, all Jewish-owned businesses were required to register with the appropriate government offices. Over the next months, Jews were fired from university and government positions, Jewish newspapers were shut down. An illegal student newspaper announced that a "cold pogrom" had begun. In January 1941, all Dutch Jews were forced to register with the state and were banned from movie theaters. This would have been a hardship for the movie-star-struck Anne, who fails to mention this as the reason for her friends being treated to a private showing of a Rin Tin Tin film at her birthday party in June 1942. By then, the basic necessities of life—food, transportation, shelter, safety—had become problematic for Jewish families. Nonetheless, the Franks sent cookies to school with Anne so she could celebrate her birthday there as well, and found a way to entertain "lots of boys and girls."

Otto and Edith Frank did everything they could to regularize their family's increasingly restricted daily existence. If the children couldn't go to the theater, they could watch Rin Tin Tin at home. On her first visit to the Franks' apartment, Miep Gies observed toys, drawings, and other signs that the children "dominated the house." Later, she noticed that the adults' conversations about the war ended abruptly when the girls appeared, and didn't resume until after the girls had finished their cake and left the room.

When Anne describes the fun she has with her Ping-Pong club at the only ice cream parlors to which Jews were still permitted to go, she doesn't allude to the incident at Koco, a similar establishment that was the scene of a battle between German police and Jewish customers. The incident led to the execution of Koco's German-Jewish owner, who refused, under torture, to name the person whose idea it was to rig up a device that sprayed the Nazis with ammonia.

Jews were forbidden to give blood, to sit on park benches, to attend horse races, or to travel. The Nazi newspaper celebrated the exclusion of Jews from the country's beaches: "Our North Sea will no longer serve to rinse down fat Jewish bodies."

As the political violence increased, Dutch citizens began to witness random street roundups (*razzia*) of Jews. In retaliation for a street brawl in which a Dutch Nazi was killed, four hundred young male "hostages" were arrested in a raid that began on the night of February 22, 1941, and resumed the next morning. Incensed, the unions and labor movements organized a general strike that shut down Amsterdam. The Nazis imposed a curfew, shot four strikers, and jailed twenty-two more. The strike ended two days later, or as Miep Gies recalls more sanguinely, it lasted "three marvelous days." The four hundred hostages were sent first to Buchenwald and then, as punishment for the fact that the Jewish Council had requested their release, to the labor camp at Mauthausen, where the commandant had given his son fifty Jews for target practice as a birthday present, and where nearly all four hundred hostages died.

During the spring and summer in which the ten-second film of the bride and groom was shot, Jews were forbidden to frequent parks, zoos, cafes, museums, public libraries, and auctions. No wonder so many Merwedeplein residents had nothing to do on a sunny June day but watch a newlywed couple walk down a flight of stairs.

That autumn, another law made it illegal for Jewish children to attend school with Christian classmates; all summer, the employees of the Dutch educational system had worked overtime to ensure that the segregated system was operational by the start of the fall term. A teacher at the Montessori school that Anne was forced to leave recalled that they lost eighty-seven students as a result of the new decree. Until then, they had not realized how many of their pupils were Jewish. Of those eighty-seven children, only twenty would survive the war.

Several short reminiscences collected in *Tales from the Secret Annex* concern Anne's experience at the Jewish Lyceum, to which she and Margot transferred after the ruling was put into effect. In "My First Day at the Lyceum," Anne describes being more afraid of having to take geometry than of the law requiring her to change schools. She reports having little sympathy for a gray-haired, mousy teacher, "wringing her hands" as she made organizational announcements. Perhaps her hand wringing had to do with a premonition about the Nazi educational policies, a worry which would have seemed less urgent to the children adjusting to new surroundings. Postwar research has shown that many Jewish children enjoyed their time at the Jewish Lyceum; reunions have been held, at which survivors recalled how glad they were to be at a school where they felt, however temporarily, safe.

Whatever anxiety Anne experienced was dispelled when she managed to get her friend Lies (Hanneli Goslar) moved into her class. "The school—which had given me so many advantages and so much pleasure—was now smiling down on me, and I began, my spirits soaring again, to pay attention to what the geography teacher was saying."

Other sketches portray the math teacher who called Anne Miss Quack Quack and the biology instructor whose favorite subject was reproduction, "probably because she's an old maid." In a series of answers to the questions that Anne wrote in the exercise book and titled "Do You Remember?," subtitled "Memories of My School Days at the Jewish Lyceum," the longest section is devoted to an incident in which Anne and Lies, accused of cheating on a French test, explained that the entire class was cheating, then wrote a letter of apology to their schoolmates for having snitched. The piece ends with Anne's hope that someday she will again be able to enjoy carefree school days.

The historian Jacob Presser taught history at the Jewish Lyceum. In the summer of 1942, at a ceremony to celebrate the lyceum's first anniversary, a usually reserved and circumspect colleague told Presser that the war was growing worse with every hour. Later, Presser would learn that the professor had just heard about the proposed mass deportations. As the raids and roundups became more frequent, there were often empty seats in the classroom. Presser recalled the pantomime with which he and the children acknowledged these absences. The teacher nodded at the empty desk, and the others either made a quick hand gesture, meaning *underground,* or a fist, meaning *arrested.* All this was done in silence. Hanneli Goslar remembered Presser bursting into tears during a lecture about the Renaissance; his young wife, who would be killed in the war, had been taken away the night before.

After the war, in cooperation with the Netherlands Institute for War Documentation, Presser would go on to write a definitive history of the period, *The Destruction of the Dutch Jews.* Among the remarkable aspects of Presser's book is his documentation of the bizarre respect for the legal process—the concern that everything be done according to the letter of the law—that went hand in hand with the Nazis' brutality. Each order depriving the Jews of their dignity, their liberty, and their ability to sustain themselves featured clauses and subclauses designed to make everything "clear." If a law was found to be "flawed"—for example, if it was discovered that the Jews required to hand in their radios were giving up old or damaged sets and keeping better models—a new law would be passed, modifying and improving the old one; now, upon surrendering their radios, Jews were forced to sign a statement swearing that they had not substituted inferior ones, and those who had already given up their sets were called back to fill out a declaration. When it was decreed that Jews could no longer ride in motorized vehicles, an exception was made for funerals; the corpse could be transported in a hearse, but the mourners still had to walk.

In January 1942, the Franks signed up for "voluntary emigration." And in April, the Jewish Council, established to control and pacify the Jewish population, distributed over half a million yellow stars, with directions on how they should be worn by every Jew over six years of age. The mandatory stars were given out along with a bill: each star cost a few pennies and one textile-ration coupon. One teacher at the Jewish Lyceum refused to wear his star because, he said, he refused to be led like a lamb to the slaughter. When his students argued that they had been told the star was a badge of honor—as Anne says in the dramatized versions of her diary—the teacher replied that those who thought so should wear it. Ultimately, he gave in, and his wife, weeping, sewed the star to his jacket. After a week, he decided to face the consequences and removed the star.

So many Dutch people also wore yellow stars in solidarity that, as Miep Gies reports, their South Amsterdam neighborhood was jokingly known as the Milky Way. But the Germans made it clear that this was not a joke, and after a few arrests, only the Jews were left wearing their six-pointed badges.

FURTHER along in the June 20, 1942, entry, the introductory letter that Anne added to her revisions so as to bring Kitty (and future readers of *Het Achterhuis*) up to the point at which she intends her book to begin, she lists the regulations and prohibitions that have affected her most deeply: Jews were forbidden to ride in trams. They were required to hand in their bicycles, to do their grocery shopping between three and five in the afternoon, to stay indoors from eight at night until six in the morning. They were banned from theaters and cinemas, from swimming pools and public sports, and were not allowed to visit Christians.

"So we could not do this and were forbidden to do that. But life went on in spite of it all. Jacque used to say to me, 'You're scared to do anything because it may be forbidden.' Our freedom was strictly limited. Yet things were still bearable." (The "Jacque" whom Anne refers to here is her friend Jacqueline van Maarsen.)

As if to restore her sense of perspective, Anne quickly moves on to the unbearable thing. "Granny died in January 1942." Then she returns to the subject of how the Nazi laws have affected her life, how she was forced to leave a favorite teacher when she transferred to the Jewish Lyceum.

The entry concludes, "So far everything is all right with the four of us and here I come to the present day."

So Far everything is all right with the four of us.

On December 1, 1940, almost seven months before Germany invaded Holland, the Opekta company had relocated to a new home, at 263 Prinsengracht, where, Otto told his employees, there would be room for the company to grow. The business was doing well, especially after Hermann van Pels—a friend of Otto's who had run a meat-seasoning company before he too left Germany for Holland—was brought in to oversee the subbranch, Pectacon, trading in spices used in sausage making and pickling. This enabled Opekta, whose jam-making products were in demand only in the summer and autumn, to turn a profit year-round. Otto Frank and Hermann van Pels worked together, lived near each other, and as the Nazis' plans for the Jews emerged, made plans to go into hiding, with their families, in the annex behind the office. Van Pels introduced Miep Gies to a friendly butcher who would later provide meat for the hidden Jews.

In January 1942, fifteen senior Nazi officials, including Adolf Eichmann and Reynhard Heydrich, head of the Reich Security Office, met in the Berlin lakeshore suburb of Wannsee to draft "the final solution to the Jewish question." Over half a million German and Austrian Jews had already emigrated since 1933, and, at the Wannsee Conference, Heydrich offered this ingenious and ambitious plan for disposing of those who remained in Europe:

> Under proper guidance, in the course of the final solution the Jews are to be allocated for appropriate labor in the East. Able-bodied Jews, separated according to sex, will be taken in large work columns to these areas for work on roads, in the course of which action doubtless a large portion will be eliminated by natural causes. The possible final remnant will, since it will undoubtedly consist of the most resistant portion, have to be treated accordingly, because it is the product of natural selection and would, if released, act as the seed of a new Jewish revival.

The Wannsee meeting was challenging but productive, and afterward, Eichmann, who would be responsible for implementing the new protocols, described his colleagues enjoying a much-deserved opportunity to relax. "At the end, Heydrich was smoking and drinking brandy in a corner near a stove. We all sat together like comrades . . . not to talk shop, but to rest after long hours of work." During this leisurely chat, Eichmann and Heydrich hashed out the details of how "the final solution" would be put into practice.

As early as 1938, Otto had applied in Rotterdam for a visa that would allow his family to emigrate to the United States. But by the next year, 300,000 applicants were on the waiting list for visas. According to the letters discovered at YIVO in 2007, Otto began writing, in April 1941, to his college friend Nathan Straus, the Macy's department store heir who was then serving as head of the U.S. Housing Authority, a New Deal agency.

Dignified and polite, remarkably restrained in view of the increasing desperation of his situation, Otto asked for the financial and political help that would allow the Franks to leave Holland. He apologized for imposing and assured his former schoolmate that he would not be bothering him if not for the sake of his children. Written between April and December 1941, these letters failed despite the support of Edith's two brothers, who lived in Massachusetts and were willing to sponsor the family and underwrite their passage.

In the end, U.S. immigration policy proved too inflexible to bend even under pressure from Straus. When Otto explored the possibility of emigrating to Cuba, he was granted a visa on December 1, 1941. But the visa was canceled when, a few days later, the United States declared war on the Axis powers. In January 1942, Otto again applied for permission to leave Holland. But by then, such applications could only be submitted to the

Jewish Council, which was unable to arrange for emigration to anywhere except Westerbork—and Poland.

On June 20, 1942—the date Anne put on the entry in which she wrote about how "odd" it was that an ordinary girl like herself should keep a diary, and named her little book Kitty—Adolf Eichmann and Franz Rademacher, a virulently anti-Semitic official at the Foreign Affairs Ministry in Berlin, agreed that 40,000 Dutch Jews should be sent to Auschwitz. After much negotiation, the Jewish Council agreed to come up with 350 names per day, and it was decided that the deportations would begin on July 5. By the end of July, 6,000 Jews had been deported.

Among the European countries under Nazi control, Holland lost, second only to Poland, the largest percentage of its Jews; more than three-quarters of Dutch Jews were killed. Several factors contributed to Holland's dismal record. The Netherlands was bordered by occupied territory, making escape more difficult. The terrain lacked woodlands and underpopulated areas in which to hide. The capture of the Jews was facilitated by hyperefficient Dutch record keeping, which made it easy for the Germans to find them, and by the initial and persistent disbelief of the Dutch and the Dutch Jews.

For every Resistance worker and brave Dutch citizen who risked imprisonment or death to hide imperiled Jews, others were unable or unwilling to help, and in fact did whatever was necessary to placate the Germans. Municipal clerks stamped J's on identity documents, impounded Jewish radios and Jewish bicycles, and sent the Jewish unemployed to labor camps. Dutch workers made sure that the commandeered bicycles were in perfect shape, and were equipped with spare tubes and tires provided by the Jews giving them up. According to one Dutch civil servant, "Often one made an effort to be ahead of the Germans, in order to do what one supposed the Germans would do, at least what one supposed the Germans would like."

Who can say, with conviction, what he or she would have done in their place? The Dutch people knew that their safety and livelihood and the survival of their families was at stake. "Everybody had a family to support: *the sense of responsibility towards the family was never greater than during the years of the occupation*," was the bitter observation of one Dutch Resistance hero, Henk von Randwijk.

Given that there were never more than 200 German policemen in Amsterdam, the majority of the raids and arrests were performed by Dutch police, and by civilians paid a bounty for turning in Jews. From July 1942 to September 1944, 107,000 Jews were deported. According to Adolf Eichmann, the Dutch transports ran so smoothly that they "were a pleasure to behold."

ON THE other hand, there was Miep Gies. Originally Hermine Santrouschitz, an Austrian Christian who had come to Holland as a child to escape post–World War I food shortages and was subsequently adopted by her Dutch foster family, Miep had been given a Dutch name and thought of herself as Dutch.

As a student, she was interested in philosophy and literature. Like Anne, she kept a notebook. But unlike Anne, Miep abandoned her dreams of writing, left school, and got an office job. She was unemployed when, in 1933, a neighbor who worked as a traveling sales representative for Otto Frank's company told her about a vacancy in the Opekta office.

Miep and Otto Frank liked each other at once. After proving that she could master the intricacies of jam and jelly making, Miep was hired as a sort of one-woman complaint bureau to help customers who called to report home-canning problems. Miep was promoted, until her job combined the duties of a secretary, an office manager, and an assistant to Otto Frank.

Miep and her fiancé, Jan Gies, were often invited to the

Franks' Saturday afternoon open houses, where Otto Frank introduced them as his Dutch friends. The Franks' widening social circle would grow to include Hermann van Pels, his wife, Auguste, and their son, Peter. Miep also met Fritz Pfeffer, the dentist who would appear in Anne's diary as Dussel, and whom Miep remembered as having the dashing charm of Maurice Chevalier. Pfeffer became her dentist, and she remained his patient even after Christians were forbidden to consult Jewish health professionals.

In 1938, Miep was shocked when she went to renew her Austrian passport and it was seized and replaced with a German one. Soon after, she was recruited by a "very blond young woman" and asked to join the Nazi party, an invitation Miep turned down, she told the recruiter, because of the Nazis' treatment of the Jews. After the invasion and the February strike, the Germans revoked her passport, and she was told she would have to go back to Vienna unless she joined the Nazi party or married a Dutchman.

Miep and Jan—a Dutch-born social worker with whom she would share the work and danger of hiding eight people above a functioning business—had already decided to marry; Hitler hurried along their engagement. A panic over the location of her birth certificate, required for the wedding, was solved thanks to the intercession of a relative in Austria.

A group photo of the Frank family shows them—with the exception of Edith, who had stayed home with her mother, who was very ill—in party clothes and in remarkably (given the escalating Nazi regulations) high spirits, on their way to Miep's wedding on July 16, 1941. It was almost a month after the camera caught Anne watching another bridal couple from her Merwedeplein window. The occupation was already thirteen months old; a wedding must have seemed like a welcome distraction.

In her memoir and on film, Miep Gies gives the impression of being one of those rare people for whom independence, conscience, and the impulse to do the right thing are matters of reflex as much as choice. "Jews were such an established part of the fabric of city life," she writes, "there was nothing unusual about them. It was simply unjust for Hitler to make special laws about them."

After the passage of the law mandating the wearing of yellow stars, and after the ban on Jews taking public transportation meant that Otto, already middle aged, had to walk the long distance back and forth to work, Otto asked Miep if she would help his family go into hiding.

She didn't ask, as anyone reasonably might, for time to think it over.

"There is a look between two people once or twice in a lifetime that cannot be described by words. That looked passed between us." Otto Frank reminded Miep that she could be sent to prison for helping to conceal them, but she already knew that. Miep and Jan, who worked for the Resistance, also found hiding places for their Jewish landlady and her two grandchildren, another good deed that would have unexpectedly fortunate consequences for Otto Frank and the others after the secret annex residents were sent to Auschwitz.

For security reasons, no one in the Prinsengracht attic was informed when Miep and Jan Gies hid a young man in their own home, a Dutch student who had refused to sign an oath pledging that he would not take action against the Germans, and whose adolescent rebelliousness made him a dangerous guest. Miep and Jan knew people who knew people who knew how to survive and get things done, how to forge ration cards and buy extra sugar. Such information was available if people thought you could be trusted.

THE IDEA of the attic hiding place had originated with Johannes Kleiman, the Opekta bookkeeper and a company board member. He and Otto Frank had been friends for almost twenty years, since Kleiman was associated with the Frank family bank that Otto tried to establish in Amsterdam. When the Dutch branch failed, Kleiman let Otto use his home address as that of the bank so that Otto could avoid paying rent on a commercial building until he settled his debts. Kleiman was helpful again when Nazi regulations prohibited Otto from owning a business. The Opekta stock was registered in Kleiman's name, and he assumed official control of the firm, though every major decision was still overseen by Otto. Kleiman mentioned the attic as a possible refuge to Otto Frank in the summer of 1941, six months after the company moved into its new quarters at 263 Prinsengracht, and a year before the Franks went into hiding.

Kleiman was married and had a child. He was frequently ill with severe stomach problems that worsened during the war and from which he never recovered. He was said to have had a soothing presence. Anne, whose pseudonym for Kleiman was Koophuis, claimed that his visits always cheered her up and quotes her mother as saying, "When Mr. Koophuis enters, the sun begins to shine." In the diary, we see him scattering flea powder, investigating a burglary, getting bread from a friend who worked as a baker. Thin, bespectacled, awkward, he appears, in a postwar photograph, hovering uncomfortably in the background with Otto's second wife, Fritzi, as Otto stands in front of 263 Prinsengracht with the playwrights and the director who would soon bring Anne's diary to Broadway.

The top-floor rooms had been used as a laboratory by a Mr. Lewin, Anne's pseudonym for Lewinsohn. There, the pharmacist-chemist, a friend of Otto's, had experimented, concocting hand creams and such. The space was vacated when the anti-Jewish laws prevented Lewinsohn from practicing even his humble brand of science. Later, Anne would dread Mr. Lewinsohn's visits to the downstairs office; she feared that he might decide to "have a peep in the old laboratory."

Together with Viktor Kugler, an Austrian-born, national-ized Dutch citizen who was married to a Dutch woman and who had worked for Otto Frank since the early days of Opekta, Kleiman—with the help of his brother—prepared the annex, installing a bathroom, doing the plumbing and carpentry, and amassing the provisions required to make the place habitable. They moved furniture, cooking equipment, and food supplies to the office after it became illegal for Jews to transport house-hold goods and to go out at night, when most of the work had to be done so as to keep it secret. Among the things we rarely consider about life in the secret annex is the sheer *amount* of food—the sacks of potatoes and beans, the quantities of meat and milk—required to feed eight people, three meals a day, for two years and one month. Much of that responsibility and the hard labor would fall to Miep Gies, whom Anne describes as having so much to carry that she looked like a pack mule.

In our imaginations, the annex is simply *there*. In the Broad-way play, the stage has been set, awaiting the Franks' entrance. And that is essentially how it appears in Anne's diary. Not until she and her parents have fled their apartment forever do the adults tell her where they are going. Of course, they have been protecting her from the danger of knowing too much, and she is soon given a chance to demonstrate how she has inherited, or absorbed, that thoughtful, protective impulse. A curious girl, Anne would have known exactly how the annex had been pre-pared, but the diary is unforthcoming with the details of the renovation, and the identity of those responsible.

In her first draft Anne writes, "We would be going to Daddy's office and over it a floor had been made ready for us."

This draft includes an astonishing inventory of *stuff* (divans, tables, bookshelves, a built-in cupboard) and one detail in particular —"there were 150 cans of vegetables and all sorts of other supplies"—that one misses in the revision. Not until she includes the floor plan of the annex, which makes it clear how much work must have been involved, does she give a nod—and not a terribly incriminating one—to the Dutch helpers. And only in the revision do we learn that "Mr. Kugler, Kleiman and Miep, and Bep Voskuijl, a twenty-three-year-old typist . . . all knew of our arrival."

To avoid endangering more people than necessary, Miep and Jan Gies were not informed until the attic was almost ready. It must have taken great care, planning, and ingenuity to keep the intelligent, observant Miep—who worked long hours at the office—from suspecting that heavy construction was going on when the firm was closed.

Once the Franks had gone into hiding, it was Johannes Kleiman who kept up a coded correspondence with Otto's family in Basel, Kleiman to whom Anne showed her stories and whom she begged (he refused, because it seemed too dangerous) to submit her essays and tales to a newspaper, Kleiman who found medicine for Margot's bronchitis, Kleiman who engaged his contractor brother to repair the roof of the annex after it was damaged by a storm. When the building was sold, it was Kleiman who thought quickly and pretended to have lost the key that would have admitted the new owner to the secret annex.

Kleiman's daughter Corrie knew Anne. When his wife visited the secret annex, Anne would pester her for news of Corrie and of the world of teenagers outside. Once, by accident, Kleiman mentioned Otto Frank within earshot of his daughter.

"A few days later, at the table Corrie was telling a story about school. She stumbled over and mispronounced a difficult name, and I corrected her. Suddenly she looked at me and said: 'You sometimes get names twisted, don't you?'

"After that she never said another word about it. But now she knew, and kept her silence. Children can be very loyal, to themselves and others, and Corrie was deeply attached to Anne."

Like Kleiman, Viktor Kugler—who became the proxy director of the firm, renamed Gies and Company, after Otto was no longer allowed to run it—played a crucial role in helping the Franks. It was Kugler's idea to construct the bookcase that hid the entrance to the annex. While the Jews were in hiding, Kugler (who appears in the diary as Kraler) visited the annex hiding place almost daily, bringing magazines, newspapers, and other necessities and trying to maintain the group's morale by being optimistic and by withholding bad news. He also bought black-market ration coupons.

"Primarily, however," Otto recalled, "Mr. Kugler sold bulk orders of spices without recording the sale and he was then able to pass the money along to us. This was terribly important because over the course of the two-year hiding period our supplies were being used up. The responsibility that Mr. Kugler took upon himself was an enormous burden and he was always stressed, especially since his wife knew nothing about us, so he couldn't talk to her about his worries."

THE ANNEX was still being prepared for its residents to arrive on the target date of July 16 when, around three in the afternoon on Sunday, July 5, 1942, a messenger arrived at the Franks' door with a registered letter. It was not, as they had expected and feared, a work summons for Otto, but rather for sixteen-year-old Margot. A list of demonically precise instructions—exactly what she was required to bring and how it should be packaged and labeled—accompanied the order directing her to report to Central Station for transport to Westerbork.

Otto was away from home, visiting inmates at the Jewish Hospital, the Joodse Invalide, which Crown Princess Juliana had toured in the late 1930s as a protest against Nazi anti-Semitism, and which the Nazis would empty of its patients later in 1942. When Otto returned, he reassured his family that they would go into hiding the next day, leaving behind evidence to suggest that they had escaped to Switzerland. That was what Hanneli Goslar was told when she went to look for Anne: they were safe in Switzerland. She would have no reason to doubt this until she and Anne met in Bergen-Belsen, shortly before Anne's death.

On the morning after Margot's summons arrived, Margot and Miep bicycled through the rain to 263 Prinsengracht. Wearing as many layers of clothing as possible, trying to appear unhurried, Anne and her parents took the long walk from Merwedeplein to the city center. When they arrived in the annex, the reality of their new lives, and of the danger they had so narrowly averted, paralyzed Margot and Edith, leaving Otto and Anne to fix up the attic. "The whole day long, we unpacked boxes, filled cupboards, hammered and tidied, until we were dead beat. We sank into clean beds that night." Such sentences are typical of the way in which Anne managed to make an insane and horrifying reality—a family was about to spend two years in an attic to avoid being rounded up and killed—seem (as her parents must have wished it to seem) merely like an unusual turn in the normal course of events.

A week later, the Van Pelses—Hermann, Auguste, and Peter —arrived, bringing news of how dangerous life had become in just a few days. The frequency and violence of the arrests had increased, people were being dragged from their houses and taken to a theater, the Hollandsche Schouwburg. Jews were crowded into streetcars, transported to Central Station, and sent to Westerbork.

By the following autumn, Jewish homes were being raided nightly. In November, the month Miep's dentist arrived to become the eighth resident, two thousand Jews were shipped to Westerbork.

Anne wrote in her diary, "In the evenings when it's dark, I often see rows of good, innocent people accompanied by crying children, walking on and on . . . bullied and knocked about until they almost drop. Nobody is spared, old people, babies, expectant mothers, the sick—each and all join in the march of death. . . . And all because they are Jews!"

SOONER or later, experience teaches us how circumstances distort our perception of time. How rapidly the hours pass in the presence of a loved one, how slowly the seconds crawl by when we are stalled in traffic. It is hard to imagine how the residents of the annex got through the long days during which they were forbidden to move or cough or flush the toilet, the hours marked off, in fifteen-minute intervals, by the bells in the Westertoren. Though her mother and sister found the bells maddening, their tolling was, to Anne, "a faithful friend."

Divided into quarter hours, two years and one month passed until, on August 4, 1944, the annex was raided and its occupants arrested.

Over the intervening decades, considerable interest has surrounded the identity of the person who betrayed the Franks. Later, the helpers agreed that someone must have turned in the Jews. Much of the suspicion has fallen on a warehouse worker named W. G. van Maaren, who was hired when Johannes Voskuijl—a trusted employee who built the bookcase that concealed the secret annex—became too ill to work. The helpers communicated their mistrust of Van Maaren to the Franks.

"Another thing which doesn't cheer us up is the fact that the warehouseman, v. Maaren, is becoming suspicious about the Annex. Of course anyone with any brains at all must have noticed that Miep keeps saying she's off to the laboratory, Bep to look at the records, Kleiman to the Opekta storeroom, while Kugler makes out that the 'Secret Annex' is not part of our premises but belongs to the neighbor's building."

Van Maaren's suspicions grew, and became even more disquieting, after he found the wallet that Hermann van Pels had accidentally left downstairs in the office one night. Johannes Voskuijl's daughter, Elizabeth "Bep" Voskuijl, originally a secretary who became an administrator at Opekta and one of the Franks' helpers, recalled Van Maaren noticing all the little slip-ups and mistakes the annex residents made, the pencils left out on a desk, a cat's drinking bowl filled during the night. Bep was outraged when it was Van Maaren who brought her and Miep a sheaf of Anne's papers that he had salvaged from the attic after the Franks were arrested.

A series of postwar investigations into Van Maaren's alleged perfidy centered on precisely what words or signals had passed between him and the police who arrived on August 4 and asked where the Jews were hidden. But in 1964, the last of the inquiries into Van Maaren's wartime behavior was terminated for lack of "concrete results." Other suspects have included the wife of one of Van Maaren's assistants, a man named Lammert Hartog, who told his wife that he had seen large quantities of food delivered to the warehouse, and a Dutch Nazi said to have blackmailed Otto Frank for expressing anti-Nazi sentiments.

Though he himself suggested that Van Maaren was the likeliest culprit, Otto chose not to focus his energies on bringing his family's betrayer to justice. By nature, and as a consequence of his tragic experience, he favored reconciliation over retribution, mercy over justice. What did it matter who made the telephone call that the arresting officer reported receiving? How would it have helped Otto Frank, or his wife and children, or their Dutch helpers, to know who had sold eight lives in exchange for the bounty that the Nazis paid for the fugitive Jews?

AFTER the arrest, the prisoners were taken to the Gestapo headquarters on Euterpestraat, then to the prison, the Huis van Bewaring, on Weteringschans, where they spent three nights. Otto Frank was interrogated concerning the whereabouts of other hidden Jews, but he insisted that, after twenty-five months in the annex, he had no contact with anyone who might be in a situation like his—a claim so clearly logical that even the police were persuaded.

On August 8, the eight Jews from the annex were sent by train to Westerbork. For much of the journey, Anne stared out the window at the summer day and at the world she had left behind two years and one month before. On their arrival at Westerbork, they were classified as "criminal Jews"—Jews who had gone into hiding or had otherwise refused to be "voluntarily" deported. They were issued special uniforms (blue overalls, a red bib, ill-fitting wooden shoes) and sent to the punishment barracks.

Westerbork's parody of normal life included well-equipped hospitals staffed by excellent doctors who treated prisoners so that they could be sent off to their deaths. There were entertainers, a cabaret, a symphony orchestra, soccer games—all under the guard towers and in the sights of the SS machine guns. A musical revue was performed on Tuesday evenings; earlier on Tuesdays, the weekly transports left Westerbork for Auschwitz.

In the journal she kept throughout the occupation and continued to write in at Westerbork, Elly Hillesum, a young Dutch Jewish woman a decade or so older than Anne, describes babies dragged from their cots and pregnant women forced onto the transport. One young boy tried to run away when he realized he was going to Poland. As a deterrent to keeping other children from panicking at the last moment, fifty additional prisoners were added to those who had been scheduled to go with him. "Will the boy be able to live with himself, once it dawns on him exactly what he's been the cause of?" wrote Hillesum. "And how will all the other Jews on board the train react to him? That boy is going to have a very hard time."

In Willy Lindwer's documentary, *The Last Seven Months of Anne Frank*, and in the book made from the film, the women who met the Franks at Westerbork recalled seeing the family together. Otto asked Rachel van Amerongen-Frankfoorder, who was assigned to scrub toilets and distribute overalls and clogs to new arrivals, if Anne could work with her, but that proved impossible to arrange. Janny Brandes-Brilleslijper, a Jewish woman who had worked with the Dutch Resistance and who later served as a nurse at Bergen-Belsen, recalled that the Frank girls were among the women who worked scouring the insides of batteries:

> That was very messy work, and no one could understand the reason for it. We had to chop open the batteries with a chisel and a hammer and then throw the tar in one basket and the carbon bars, which we had to remove, into another basket; we had to take off the metal caps with a screwdriver, and they went into a third basket. In addition to getting terribly dirty from the work, we all began to cough because it gave off a certain kind of dust.

Another Westerbork prisoner remembered that Anne and Peter were always together, and that Anne, frail and extremely pale at first, came to seem radiant, even happy. Edith Frank appeared stunned and mute; Margot rarely spoke.

Ronnie Goldstein-van Cleef rode with the Frank family on the train from Westerbork to Auschwitz on September 3, 1944, the last transport that would take that route. The Franks agreed that, if they survived, they would try to find each other through Otto's mother in Basel.

More than a thousand people were on the train. The list of passengers included the Franks, the Van Pelses, and the dentist Fritz Pfeffer, all packed into boxcars. They traveled for three days and two nights before reaching Auschwitz-Birkenau, where, as was customary, they were "selected" according to their age and their level of health and fitness. Over five hundred of the new arrivals went straight to the gas chambers. None of the former residents of the secret annex was among them.

According to camp protocol, men and women were separated, stripped, and sent to be "disinfected." The hair that Anne was so proud of—and which, according to her girlhood friend, had kept her so busy—was shaved off. Numbers were tattooed on the prisoners' forearms. Anne, Margot, and their mother remained together in Women's Block 29.

Janny Brandes-Brilleslijper recalls female *Kapos* in angora sweaters following the prisoners around with whips. Recently donated to the United States Holocaust Memorial, a photo album that had once belonged to an adjutant to the Auschwitz commandant shows the guards eating blueberries, lighting a Christmas tree, relaxing in lounge chairs. Meanwhile, the prisoners worked in weaving mills and prepared plastic to be used in airplanes, or labored at more useless and punitive tasks, digging up stones and patches of sod.

Badly infected with scabies, covered with sores, Anne was sent to a particularly dreadful compound, the so-called scabies block, the barracks for prisoners with contagious skin diseases. Margot went along voluntarily, so as to remain with her sister. At Auschwitz, the debilitating anxieties that had plagued Edith in the secret annex vanished, as did the muteness and near paralysis from which she had suffered at Westerbork. Seized by the determination to help her daughters survive, Edith tunneled under the wall of the scabies block, digging a hole through which she was able to pass them a piece of bread.

In October 1944, as the Russian army approached Auschwitz, eight thousand women, including Anne and Margot Frank, and Auguste van Pels, were transferred to Bergen-Belsen, inside Germany and farther from the encroaching enemy. Edith Frank was left in Auschwitz, where the gas chambers and crematoriums were blown up to destroy the most damning evidence of what had transpired there. Edith Frank died of disease and exhaustion a few weeks before the camp was liberated by the Russians at the end of January 1945.

The Frank sisters remained together in Bergen-Belsen. The most recent paperback edition of the diary, the one students read, concludes, as do previous editions, with an afterword adapted from Ernst Schnabel's 1958 *Anne Frank, A Portrait in Courage*. Schnabel describes Anne's deportation, her imprisonment in Auschwitz, "a fantastically well-organized, spick-and-span hell," and her transport to the filthy, chaotic Bergen-Belsen:

"'Anne, who was already sick at the time,' recalled a survivor, 'was not informed of her sister's death, but after a few days she sensed it, and soon afterwards she died, peacefully, feeling that nothing bad was happening to her.' She was not yet sixteen."

In subsequent years, other reports have painted a more harrowing picture of Anne's final days.

Originally one of the "better" camps, Bergen-Belsen had degenerated into a hell of chaos and squalor. Tents were erected to shelter the new arrivals at the overcrowded compound, but a storm knocked down the makeshift housing, wounding many of the prisoners, killing some, and leaving the rest unprotected from the cold rain and hail.

At Bergen-Belsen, Anne met Hanneli Goslar, who described her as "a broken girl." Hannah, the "Lies" whom Anne had seen in a vision, desperate and starving, had in fact been spared the horrors of Auschwitz and had gone directly from Westerbork to a slightly less terrible section of Bergen-Belsen, in part because her family had Paraguayan passports, which they had purchased in Holland.

At the Anne Frank Museum, a video monitor plays and replays a filmed interview in which, after fifty years, Hannah Pick-Goslar remembers Anne weeping as she said that she no longer had any parents.

"I always think," Hannah reflects, "if Anne had known her father was still alive, she might have had more strength to survive." Hannah arranged to throw a small package ("a half a cookie, a sock, a glove") over the fence, at night. The first time, another prisoner stole the package. Hannah describes Anne screaming when the package was stolen. A second package reached her—but failed to prevent the inevitable.

In *The Last Seven Months of Anne Frank*, Rachel van Amerogen-Frankfoorder reports that the emaciated Frank girls "had little squabbles, caused by their illness . . . They were terribly cold. They had the least desirable place in the barracks, below, near the door, which was constantly opened and closed. You heard them constantly screaming, 'Close the door, close the door,' and the voices became weaker every day . . . What was so sad, of course, was that these children were so young . . . They showed the recognizable symptoms of typhus—that gradual wasting away, a sort of apathy, with occasional revivals, until they became so sick there wasn't any hope . . . One fine day, they weren't there any longer." She corrects herself. "Actually, a bad day."

Corpses were heaped near the barracks, then buried in mass graves. Rachel van Amerogen-Frankfoorder thinks she might have passed the bodies of the sisters on her way to the latrine. "I don't have a single reason for assuming that it was any different for them than for the other women with us who died at the same time."

This account, by Janny Brandes-Brilleslijper, also appears in the book adapted from Lindwer's documentary:

> Anne stood in front of me, wrapped in a blanket . . . And she told me that she had such a horror of the lice and fleas in her clothes and that she had thrown all of her clothes away. It was the middle of winter . . . I gathered up everything I could find to give her so she was dressed again. We didn't have much to eat . . . but I gave Anne some of our bread ration.
>
> Terrible things happened. Two days later, I went to look for the girls. Both of them were dead!
>
> First, Margot had fallen out of bed onto the stone floor. She couldn't get up anymore. Anne died a day later. We had lost all sense of time. It is possible that Anne lived a day longer. Three days before her death from typhus was when she had thrown away all her clothes during dreadful hallucinations.

A few weeks later, Bergen-Belsen was liberated by the British.

THE play based on *The Diary of Anne Frank* begins and ends with scenes of Otto Frank returning after the war, telling Miep Gies that his wife and daughters are dead, and finding Anne's journal in the wreckage of the annex. No one asks or explains what happened to the others with whom the Franks hid, to the real men and woman on whom the playwrights modeled the characters who have entertained, maddened, and moved us in the production.

In Auschwitz, the four men from the attic were sent to the same barracks. There they happened to meet Max Stoppelman, a tough, broad-shouldered Dutch Jew in his thirties, hardened by—and experienced in—the way things worked in the camp. His mother had been Miep and Jan Gies's landlady in Amsterdam, and the couple had arranged for her and other family members to go into hiding. Stoppelman was grateful to the Franks, by association. So it could be said that Miep and Jan were still helping the Franks, even from a great distance.

Stoppelman took a liking to the four men, especially Peter. Under his tutelage and protection, Peter managed to get a job in the camp post office. Less could be done for the older men—better candidates, according to Nazi logic, for brutal outdoor labor. Hermann van Pels was gassed in the fall of 1944 after a finger injury sapped his ability to work and his will to survive.

Emaciated and exhausted, Otto wound up in the sick bay, where he was watched over by Peter van Pels, who took filial care of the three men with whom he had shared the annex. In October, Peter was evacuated to Mauthausen, where he died in May 1945. A similar destiny awaited Fritz Pfeffer, moved first to Sachsenhausen and then finally to Neuengamme, where he died from illness and exhaustion in December 1944. His companion, Charlotte Kaletta, continued to wait and married him posthumously in 1950.

How one pities Auguste van Pels, so protective of her fragile dignity and her meager creature comforts, so unprepared, as if anyone could have been prepared, for what lay before her in Auschwitz, and then in Bergen-Belsen, where she was imprisoned with Anne during Anne's final days. It was Auguste who brought Anne, already ill, out to meet Hanneli Goslar. Auguste van Pels died in the spring of 1945, somewhere near Thieresienstadt, during one of the forced marches that followed the evacuation of the camps.

Down to a hundred and fourteen pounds, Otto remained in the hospital barracks as liberation drew near. At one point, the patients were ordered outside by the SS and almost shot, but the threat was rescinded at the last minute.

In January, Otto Frank was among the 7,650 prisoners whom the Russians found alive in the Auschwitz compound. It would take him six more months to return to Holland and to his daughter's diary, which had been saved by Miep Gies, who, as I write this, in late 2008, is well into her nineties and only lately in fragile health.

PART II

The Book

THREE

The Book, Part I

IN AMSTERDAM, ON THE SUNNY AND OTHERWISE QUIET morning of Friday, August 4, 1944, a car pulled up in front of the Opekta warehouse at 263 Prinsengracht.

That is all one needs to write, and already the reader knows who was hiding in the attic and the fate about to befall them. We know it more than sixty years later, at a historical moment when it is often noted how little history we remember. We know the reason why we know, but it bears repeating lest we take it for granted that we know because a little girl kept a diary. Of all the roundups, the deportations, the murders committed during that time, this arrest is the one that has been investigated most closely, the one about which memories have been most thoroughly searched. It is the one we know about, if we know about any at all.

The car arrived without sirens, without haste. Upstairs in the attic, Otto Frank was correcting Peter van Pels's English dictation. No one in the office below was alarmed by the appearance of the car until a fat man appeared, and, speaking in Dutch, ordered everyone to be quiet.

One of the men who got out of the car wore the uniform of a sergeant in the Jewish Affairs Section of the Gestapo in Holland. The officer was an Austrian, his subordinates Dutch civilians employed by the Nazis. They entered the spice and pectin warehouse, then went up to the office, where they found the Opekta staff. They demanded to know who was in charge. Viktor Kugler replied that he was. After searching the storerooms, the police pulled aside the bookcase that covered the door to the attic where, they had obviously been informed, Jews were hiding.

The first to climb the narrow, nearly vertical stairs, Kugler told Edith Frank, "The Gestapo is here." She stood still and said nothing.

The Gestapo officer and his men entered the secret annex and found the Jews, as they had expected, though they would not have known—as we do now—whom they would find. Three men, two women, a young man, a young woman, a girl.

In a few more years, no one alive will have witnessed the scene of a Nazi arresting a Jew. There have been, and will be, other arrests and executions for the crime of having been born into a particular race or religion or tribe. But the scene of Nazis hunting down Jews is unlikely to happen again, though history teaches us never to say never. This will be the arrest that future generations can visualize, like a scene in a book. They will have to remind themselves that it happened to real people, though these people have survived, and will live on, as characters in a book.

In fact this scene is not in the book, but that book's existence is the reason we know about the arrest. We know that the Austrian officer's name was Karl Josef Silberbauer. And we know that he was disturbed by the detail of Otto Frank's military trunk, labeled as the property of *Lieutenant* Otto Frank, which meant he would have been Sergeant Silberbauer's superior when both fought in the German army during World War I.

Later, Otto Frank would recall that Silberbauer seemed to snap to attention. For the Austrian, the Jew's former military rank created a troubling disruption in the simultaneously adrenalinized and business-as-usual theater of arrest.

In a photo from that period, the thirty-three-year-old Silberbauer looks younger than his age. Posed stiffly, with slicked-back hair and a lumpy jaw, as if he has tobacco wadded in both cheeks, he wears a tie and a jacket with a tiny swastika pinned to one lapel. Miep Gies described him as looking neither cruel nor angry, but "as though he might come around tomorrow to read your gas meter or punch your streetcar ticket."

Silberbauer couldn't help asking Otto Frank how long all those people could have lived like that, crowded together in an attic behind a bookcase. He was taken aback by the answer: two years and one month. As proof, Otto Frank pointed to the doorway marked with pencil to record his daughters' growth. Look, he said, his younger daughter had already grown beyond the last mark.

A heartbreaking gesture, maybe less odd than instinctive, since his daughters, Margot and Annelies, were the center of Otto Frank's life. Those marks, which can still be seen on the wall of the Anne Frank Museum, were what he had to show for the two years in hiding. Possibly, Otto Frank imagined that the pencil lines would kindle, in the Nazi officer, a flicker of humanity.

As we know, they did not.

After the war, Silberbauer returned to Austria, where he was jailed for fourteen months on charges of having roughed up some Communists in 1938. Later he was rehired as a junior inspector on the Vienna police force. In 1963, Simon Wiesenthal tracked him down with the aid of a 1943 telephone directory that listed the names and numbers of all the Gestapo officials who had served in occupied Holland.

Guided by a hunch that Silberbauer might again be working for the Viennese police, Wiesenthal found his quarry when the official newspaper of the Austrian Communist Party reported that Silberbauer was alive and well and, indeed, a cop in Vienna. The Austrian officials launched an investigation to determine the criminality of Silberbauer's wartime activities. Otto Frank's statement—that Silberbauer had "done his duty and acted correctly," that he had been businesslike and even cordial—virtually ended the inquest, which was dropped for lack of evidence, though one might question the ruling that sending eight Jews to a concentration camp, seven of them to their deaths, would be criminal only if performed in an unprofessional manner.

Silberbauer remembered telling Otto Frank that he had a lovely daughter. He recalled her as prettier and older than she looked in the photo that, by the time Wiesenthal found him, was known all over the world. Given that Silberbauer went unpunished, it's mildly satisfying to imagine the moment he found out that the little girl he'd arrested had become a star.

The suddenly notorious Silberbauer complained to a Dutch reporter that his temporary suspension from the police force was making it hard to pay for the new furniture he'd bought on the installment plan, and that he could no longer use the pass that let him ride the streetcar for free. Asked if he had read Anne Frank's diary, Silberbauer replied that he had bought it to see if he was in it. Why did he think he might be? He knew what had happened to Anne after he flushed her out of the attic.

Did he imagine that, ill and starving, she could have kept up her diary in Auschwitz and Bergen-Belsen, pausing from her labors to record her impressions of Silberbauer?

The reporter suggested that Silberbauer could have been the first to read the book that had been read by millions. To which the Austrian replied, "I never thought of it. Maybe I should have picked it up from the floor." Wiesenthal concludes this chapter in his memoir by saying of Silberbauer, "Compared to the other names in my files, he is a nobody, a zero. But the figure before the zero was Anne Frank."

WHAT could Anne have thought when her father directed the Gestapo sergeant to the marks on the doorway? The normally attention-loving girl would doubtless have preferred that her father not focus the policeman's attention on her. Had she been able to describe this scene, and to make Silberbauer, as he seems to have wished, a character in her book, she might have recorded what came next, the detail of the briefcase, an event that would have caused her more pain and grief than anyone else in the room.

Whatever Silberbauer believed about his job, he would not have wanted to think that his military service involved state-sponsored murder and theft. The search for the valuables must have been a simultaneously uncomfortable and titillating part of his work. Naturally, there was curiosity. How much did the hidden Jews have? But it was also a matter of duty. The money seized in this way was sensibly being used to finance the Jews' own transportation to new lives, or death, in the east.

The police asked the Jews where their valuables were, and Otto indicated the cupboard where his cash box was kept. Better thieves, professionals, would have brought a bag for the stolen goods. But how would it have looked if the enforcers of Nazi justice carried duffel bags for the swag? We can only be grateful that Silberbauer, forced to improvise, grabbed a briefcase stuffed with papers.

The Jews and their Dutch helpers watched. All of them knew that the briefcase was where Anne kept her diary.

On April 9, there had been a break-in downstairs. The intruders had knocked a hole in the door before being frightened away. Afraid that the police might investigate, the families discussed what to do if capture seemed imminent. Worried that Anne's diary might be found, and that their helpers would be incriminated, the annex residents briefly considered the possibility of burning the diary. "This, and when the police rattled the cupboard door, were my worst moments; not my diary; if my diary goes, I go with it!"

A month later, Anne thought that an overturned vase of carnations had soaked through her papers. Nearly in tears, she was so upset that she began to babble in German and later had little memory of what she had said. According to Margot, she had "let fly something about 'incalculable loss. . . . '" The damage was not so severe as she'd feared, and she hung the damp sheets of paper on a clothesline to dry. *Incalculable loss* is a phrase that any writer might have used in response to the possibility that a manuscript could have been ruined, yet another indication of how seriously Anne took her work.

Eventually, it was decided that the briefcase containing the diary would be among the things the family took with them if a fire or some other emergency necessitated a hasty escape from the attic. But now the briefcase was being put to a different use. Silberbauer dumped out the papers, along with some notebooks, and handed the satchel to his colleagues to stuff with jewels and cash.

The detail of the briefcase could have come from one of those fairy tales that counsel reflection, patience, morality—

lest one wind up like the thoughtless, greedy man or woman (usually the wife) who mistakes the rhinestones for diamonds or cooks the magic fish for dinner. Eventually, Silberbauer realized he'd filled the briefcase with pasteboard and scattered rubies across the attic floor.

But how could he have imagined that what he had discarded—loose sheets of paper, exercise books—was not only a work of literary genius, not only a fortune in disguise, not only a record of the times in which he and its author lived, but a piece of evidence that would lead to the exposure of his role in the Nazis' war against the Jews, even as so many like him slipped back into their old lives and kept up their furniture payments?

There was no way he could have known what the briefcase contained. How could anyone have suspected that a masterpiece existed between the checked cloth covers of a young girl's diary?

ALMOST three hours elapsed between Silberbauer's arrival and that of the closed truck that transported the Jews and two of their Dutch coconspirators to the headquarters of the Security Police.

Only Miep Gies remained. For more than two years, she had brought food and supplies to the Jews and kept up their spirits by helping them maintain some semblance of contact with the outside world. Now, just when the progress of the Allies' invasion had begun to offer hope, the catastrophe they'd feared had occurred.

In Jon Blair's documentary, *Anne Frank Remembered*, Miep comes across as a sensible, dignified woman, overly modest about her English. We intuit that it would never occur to her to boast about the heroism that, for two years, was ordinary life for her and her husband. Her memoir, *Anne Frank Remem-*

bered, begins, "I am not a hero." She writes that she was only one person in the "long, long line of good Dutch people who did what I did or more—much more."

Heroic or not, the work took its toll. The bouts of illness—a gastric hemorrhage, fainting, fevers—that plagued the Franks' helpers are a recurrent motif in the diary, as is the theme of their unflagging good humor: "Never have we heard one word of the burden which we must certainly be to them, never has one of them complained of all the trouble we cause. They all come upstairs every day, talk to the men about business and politics, to the women about food and wartime difficulties, and about newspapers and books with the children." After Miep attended a party at which two policemen were among the guests, Anne wrote, "You can see that we are never far from Miep's thoughts, because she memorized the addresses of these men at once, in case anything should happen at some time or other, and good Dutchmen might come in useful." That Miep and the others were heroes is a fact that should not be overshadowed by the accusation—one of many controversies that have surrounded the diary—that the focus on the Franks' helpers has served to distract attention from the less-than-stellar record of the Dutch people's resistance to the Nazis' anti-Jewish campaign.

During one scene in *Anne Frank Remembered*, an off-screen voice, presumably the filmmaker's, reads a letter in which Otto Frank, who had died in 1980, thanks the Dutch friends who saved him. Perhaps for added drama, the voice announces that this is the first time Miep has heard the letter. It's a touching acknowledgment, but certainly this intelligent woman must have known that her former employer realized that neither he, nor his daughter's writings, could have survived without Miep.

AFTER the Jews and the two Dutch office workers were taken away, Miep found the checked diary that Anne had kept from

June until December 1942. Also scattered about were the exercise books in which she wrote subsequent volumes, the account book in which she composed the stories, essays, fairy tales, reminiscences, and novel fragments that would be collected and published as *Tales from the Secret Annex*, and finally, the hundreds of colored sheets of paper on which she had been revising the diary since the spring of 1944.

Just before he was arrested with the Jews, Johannes Kleiman told Miep Gies that it was too late to save him and the others, so she should try to save what could be salvaged from the attic. Together with Bep Voskuijl, Miep gathered up Anne's journals and the loose pages and brought them downstairs to the office. There she put them in the bottom drawer of her desk for safekeeping until, she hoped, Anne would return to reclaim them. Cannily, she left the drawer unlocked. The Nazis had had no interest in a child's papers, but, if they came back, they might wonder why someone would think a girl's diary worth locking up.

Of course, Miep was not merely guarding Anne's privacy, but protecting herself and her coworkers. Later, she would say that if she had read the diaries, she might have felt compelled to burn them, out of concern for her colleagues. It would have been safer for her to destroy the diary, just as it would have been safer not to hide eight Jews, and certainly safer for her not to go to the police headquarters on Euterpestraat on the Monday after the arrest. Even by the standards of the previous two years, Miep's attempt to bribe Silberbauer into freeing the prisoners was extraordinarily brave, an almost recklessly dangerous act that demonstrated the strength of her attachment to the Franks.

When the Opekta sales representatives were told that the Franks had been taken away, one of them (a man who was, in fact, a member of the Dutch Nazi party) took Miep aside and reminded her that the war was nearly over, the Germans were exhausted, and they would want to leave Holland with their pockets full. He himself would take up a collection from among the many friends and business associates who had been fond of Otto Frank, and Miep could go to the police and make the arresting officer, her fellow Austrian, an irresistible offer.

When Miep phoned Silberbauer, he instructed her to come on Monday, but when she did so, he told her to wait until the next day. On Tuesday, he said that he lacked the authority to make such a decision and sent her upstairs, where some officers were listening to English-language radio, an illegal act. Presumably they were monitoring the bad news about the Allies' progress. They ordered her out of the room. But in any case, there was nothing that Miep could have done. The prisoners had already been taken to the jail on Weteringschans in preparation for their transport to Westerbork.

Before the movers arrived to strip the upper floors of 263 Prinsengracht of furniture that could be shipped to needy German families whose own possessions had been lost in the war, Miep told one of the warehousemen—the same worker several helpers suspected of having betrayed the Franks—to pick up any papers still left on the floor. All of this was done in haste. There is no way of knowing if any, or how much, of Anne's writing was lost.

ALONG with Johannes Kleiman, Victor Kugler was taken to the SS headquarters on Euterpestraat. The two men were briefly interrogated, then transferred to the prison on Amstelveenseweg, and later to the Amersfoort transit camp. In mid-September, Kleiman suffered a stomach hemorrhage and was released when the Red Cross interceded on his behalf. Kugler was sent to a series of labor camps. After escaping in the spring of 1945, he remained in hiding until the war's end.

When Kleiman returned to the office, Miep refused to let

him or Bep read the diaries. The books and papers remained in her desk drawer for almost a year.

In June 1945, Otto Frank made his way back to Amsterdam. Liberated from Auschwitz, he had traveled by train to Russia and then by boat to Marseilles, and eventually reached the Netherlands by train and truck. He moved in with Miep and her husband. A month after his arrival in Amsterdam, Otto, who had already learned of his wife's death from a prisoner he'd met on the train to Odessa, was informed that his daughters were also dead—first by the Red Cross and then by a Dutch woman who had known the girls in Bergen-Belsen.

In the film, *Anne Frank Remembered*, Janny Brandes-Brilleslijper recalls giving Otto Frank the sad news:

> *He stood on the porch and rang the bell. He said, "Are you Janny Brandes?"* . . . *Because he was a very polite gentleman, he came into the hallway and remained standing there and said, "I am Otto." I could hardly speak, because it was very difficult to tell someone that his children were not alive any more. I said, "They are no more." He turned deathly pale and slumped down into a chair. I just put my arm around him.*

Like Otto, Miep had hoped that Anne might have survived. Planning to give the diary back to Anne upon her return, Miep concealed its existence from Otto until the day he learned that the girls had died at Bergen-Belsen.

The scene in which Otto Frank first read his daughter's diary is painful to imagine. Even when children have grown into thriving adults, a scrap of paper covered with childish writing can induce, in parents, a stab of nostalgia for those years that, parents may think later, were the happiest in their lives. How much more grief such pages must have occasioned when they had been written by a child so recently murdered?

In Miep Gies's memoir, she recalls that Otto Frank went to his former Opekta office and closed the door. A short time later, she brought him the papers and the checked book.

> *I could tell that he recognized the diary. He had given it to her just over three years before, on her thirteenth birthday, right before going into hiding. He touched it with the tips of his fingers. I pressed everything into his hands; then I left his office, closing the door quietly.*
>
> *Shortly afterward, the phone on my desk rang. It was Mr. Frank's voice. "Miep, please see to it that I'm not disturbed," he said.*
>
> *"I've already done that," I replied.*

Over the next months, Otto Frank tried, with little success, to reestablish himself in business. Meanwhile, he sorted through the diary, and the exercise and account books in which Anne had recorded her secrets and in which, he would say later, he met a daughter he had never really known. The same words that had consoled Anne may have provided some comfort for her father, or at least distraction, as he typed up sections of the manuscript, translated them into German, and sent them to his mother in Switzerland.

Relying mainly on the draft that Anne had revised but also borrowing from the original version and from the stories and occasional pieces that would appear as *Tales from the Secret Annex*, Otto Frank typed up a second and much longer manuscript of the diary. By now he was persuaded that Anne had wanted her diary to be published as a book entitled *Het Achterhuis*. Once the idea occurred to him, he was obliged to read the diary as something other than a personal record and with a view to how it might be read by strangers. Otto changed the names, as Anne had directed. But he kept his own family's real names, though Anne had wished the Franks to become the Robins: Frederik and Nora, and their daughters, Betty and Anne. He retained the first name of Peter van Pels, though Anne had specified that her young upstairs neighbor should appear in *Het Achterhuis* as Alfred van Daan.

Over the last half century, Otto Frank has been accused of prudishness, of being too ready to forgive the Germans, of censoring and deracinating Anne, of anti-Semitism, of sentimentality and cowardice, of greed and personal ambition. In fact, what seems most probable is that his editing was guided by the instincts of a bereaved father wanting to give the reader the fullest sense of what his daughter had been like. Otto cut a number of Anne's sharpest criticisms of her neighbors, either because of a desire to make her seem like a nicer person, or to protect the sensitivities of the living—for example, the dentist's sweetheart, Charlotte Kaletta, who was made unhappy enough by the passages that remained. (She was even more upset when the Broadway play of the diary portrayed her husband as a buffoon so unfamiliar with his religious heritage that the meaning of Hanukkah had to be explained to him.)

It's true that Otto chose to remove Anne's rare flashes of meanness and to tone down her impatience with smallness and hypocrisy. But if we search for the point at which her character was reduced from that of a young person with a complex, mature view of politics, history, and human nature to that of a cheerful teen, if we seek out the juncture at which her awareness that she was being made to suffer because she was a Jew became a more generalized identification with all of suffering humanity, we discover that those changes were not the result of Otto Frank's editing but rather of the ways in which the Broadway play and the Hollywood film of her diary chose to

represent its author. On stage and screen, the adorable was em-
phasized at the expense of the human, the particular was replaced by the so-called universal, and *universal* was interpreted
to mean *American*—or, in any case, *not Jewish*, since *Jewish* was
understood to signify a smaller audience, more limited earn-
ings, and, more disturbingly, subject matter that might alienate
a non-Jewish audience.

NEWLY widowed, still in mourning for his wife, Otto Frank was
understandably reluctant to see his marriage publicly judged
and found wanting by his daughter. Later, Anne's speculations
about her parents' relationship would become major news when
the "suppressed" five pages of the diary were discovered. And
yet the controversial conclusion that Anne reached—that her
parents' union was neither passionate nor romantic—was probably evident to the Franks' relatives, to their friends in Amsterdam, and to anyone knowledgeable or even curious about how
men and women behave in the presence or the absence of love
and desire. Certainly, those questions were of great interest to
Anne. In photographs of the Franks, charismatic, handsome
Otto and his relatively plain wife appear to have accidentally
wound up in the same frame. Otto's snapshots of his daughters
vastly outnumber those of Edith.

But though Otto cut Anne's most bitter references to
Edith, to his marriage, and to his wife's contentious relationship with their younger daughter, he chose not to excise
Anne's accounts of her own darkest moments. He retained
her pessimistic observations about the murderousness of her
fellow creatures, as well as her most enraged and despairing
protests against the anti-Semitism that had forced her and her
family into hiding. Likewise, he left in the passage in which
Anne asks why God has singled out the Jews to suffer. "If we
bear all this suffering, and there are still Jews left, when it is

over, then Jews, instead of being doomed, will be held up as an example."

After he had completed the preliminary editing, Otto asked a friend, the playwright Albert Cauvern, to check the manuscript for grammatical errors and mistakes in diction. The extant typescript suggests that at least someone besides Cauvern gave it a critical close reading. Also among its early readers was Kurt Baschwitz, a lecturer in psychology and journalism, who, in a letter to his daughter, called the diary "the most moving document about that time that I know" and "a literary masterpiece."

People who encountered Otto Frank during this period recall a handsome, distinguished man with the bearing and reserve of a Prussian officer—but whose eyes were perpetually red from weeping. He carried the manuscript with him wherever he went, and, at times, tears flowed down his face as he read a few pages aloud, or urged friends and strangers to read it.

The edited typescript was passed from hand to hand and across desks that included those of Jan Romein and his wife, Annie, two prominent Dutch intellectuals who thought the book should be published but were unable to convince anyone who had the power to do so. The manuscript was rejected by every editor who read it, none of whom could imagine that readers would buy the intimate diary of a teenage girl, dead in the war. In addition, the Dutch had no desire to be reminded of the suffering they had so recently endured, and, regardless of what the Dutch cultural minister in exile had promised in his radio broadcast, it was assumed that there would be little interest in a first-person account by one of the Nazis' young victims.

Luckily, the book had tenacious supporters. In April 1946, Jan Romein wrote about the diary in the daily newspaper, *Het Parool*, formerly the underground paper of the Dutch Resistance. Romein's essay, entitled "A Child's Voice," was at once impassioned and restrained. More than all the evidence presented at the Nuremberg trials, he wrote, the diary is an indictment of the "witless barbarity" of fascism, and of crimes that, only a year after the end of the war, his countrymen were already forgetting. Romein praised Anne's gifts—"an insight into the failings of human nature—her own not excepted—so infallible that it would have astonished one in an adult, let alone a child"—and eloquently described the way that literature can affect us.

"When I had finished it was nighttime, and I was astonished to find that the lights still worked, and we still had bread and tea, that I could hear no airplanes droning overhead and no pounding of army boots in the street—I had been engrossed in my reading, so carried away back to that unreal world, now almost a year behind us."

After Romein's essay appeared in *Het Parool*, Otto Frank was approached by several publishers, among them Contact, located in Amsterdam. Its managing director was interested in the book but objected to passages in which, he felt, Anne wrote too freely about sex and about her body. Otto agreed to their deletion. Perhaps he was relieved to omit those entries, which Anne herself had left out when she revised her diary. Excised from the Dutch, these sections would be reinstated when the English-language edition appeared.

In the summer of 1946, five selections from the diary were published in a journal with which the Romeins were associated. The editing continued. On the request of the Dutch publisher, roughly twenty-five pages were cut from the manuscript, including a reference to menstruation and another to Anne and a girlfriend touching each other's breasts.

Het Achterhuis was published in the Netherlands in 1947 in an edition of 1,500 copies and with a subtitle that read: *Diary Letters June 12, 1942–August 1, 1944.* Annie Romein's introduction featured a somewhat more moderate estimation of the book's merits than the rave, by her husband, that helped arrange its publication. Her preface was the first of many responses to the diary that praised the book while dismissing it as the unaffected, unpolished scribblings of an unusually gifted child.

This "diary of a normal child growing up in exceptional circumstances . . . will disappoint anyone who hopes to experience a 'wonder '. . . this diary is not the work of a prodigy." The diary "is not the work of a great writer, but the awakening of a human soul is drawn so purely, precisely, and uncompromisingly that one seldom sees the like in the memoirs of the greatest writers." Like the children in Richard Hughes's *A High Wind in Jamaica,* wrote Romein, Anne Frank sees the world and especially herself "with a candid frankness, unjudgemental and down to earth."

Without knowing that she was weighing in on a soon-to-be controversial question—the issue of whether or not Anne's diary was, strictly speaking, a Holocaust document—Romein observed, "This diary is also a document about the war, about the persecution of the Jews. The life of those in hiding is beautifully described by this child who had in any case that one essential quality of a great writer: to remain unbiased, to be unable to get used to, and therefore blinded by, the way things are."

The other thing Romein seems not to have known was how carefully Anne revised her diary, nor does she seem to have taken seriously Anne's reflection, in the aftermath of the Dutch minister's radio broadcast, on how interesting it would be if the romance of *Het Achterhuis* were published.

"The diary is pure conversation with herself. There's not

one disturbing thought about future readers, not one faint echo of . . . the will to please."

Favorably reviewed, Anne's book did moderately, if not extremely, well in the Netherlands. It was reprinted again at the end of the year, twice in 1948, once in 1949, and not again until 1950—after which it went out of print until 1955, when its success in the United States created new demand. However modest, the book's reception in Holland helped interest editors elsewhere in Europe. The Dutch edition was in its sixth printing when *Das Tagebuch der Anne Frank* was published in Germany in 1950.

In an attempt to capture Anne's voice, the German translation made mistakes in tone, and, for fear of alienating its projected audience, omitted references to the anti-German sentiment in the secret annex. The prohibitions against listening to German radio stations and speaking German—a problem for Mrs. Frank and Mrs. Van Pels, who had never become entirely fluent in Dutch—appear nowhere in *Das Tagebuch*, and a reference to the hatred between Jews and Germans was changed to read "*these* Germans."

In an April 1959 interview in *Der Spiegel*, the original Dutch-into-German translator, Anneliese Schütz, explained. "A book intended after all for sale in Germany cannot abuse the Germans." This reluctance to offend readers in a country whose leaders had murdered the book's author was one gauge of the speed at which the diary had already become a commodity that the public might, or might not, choose to buy. Despite the editing changes, the first printing was not a commercial success in Germany.

IN THE United States, Anne Frank's diary was initially rejected by nearly every major publishing house. "It is an interesting document," admitted an editor in the American branch of the international firm Querido, based in Amsterdam, "but I do not believe there will be enough interest in the subject in this country to make publication over here a profitable business."

Ernst Kuhn, a friend of Otto's who worked at the Manufacturers Hanover Bank in New York, took on the challenge of trying to find the diary an American home. Just as in Europe, the book was viewed as being too narrowly focused, too domestic, too Jewish, too boring, and, above all, too likely to remind readers of what they wished to forget. Americans did not want to hear about the war. "Under the present frame of mind of the American public," an editor at Vanguard wrote Kuhn, "you cannot publish a book with war as a background."

Alfred A. Knopf, Inc. turned down the manuscript on the grounds that it was "very dull," a "dreary record of typical family bickering, petty annoyances and adolescent emotions." Sales would be small because the main characters were neither familiar to Americans nor especially appealing. "Even if the work had come to light five years ago, when the subject was timely . . . I don't see that there would have been a chance for it." While recognizing that "so few contemporary books or documents (are) as genuine or spontaneous as this one," Viking decided that it was an infelicitous moment for the diary to appear. "If times were normal I would do an edition and translation," wrote one editor, "but times are not normal."

In Great Britain, the reaction was similar. At Secker and Warburg, it was felt that "The English reading public would avert their eyes from so painful a story which would bring back to them all the evil events that occurred during the war." As proof, an editor there noted that *The Wall*, John Hersey's novel about the Warsaw ghetto uprising, was "not doing as well as expected."

In one of her "Letters from Paris" that appeared in the
New Yorker, Janet Flanner referred to the current popularity of
a book by "a precocious, talented little Frankfurt Jewess." Yet
despite the *New Yorker* mention, and despite the book's reception in France, Anne's diary was in the reject pile in the office of
Frank Price, the director of Doubleday's foreign bureau, when
a young assistant named Judith Jones—who would go on to
become a legendary editor at Knopf, working with authors including Julia Child—found it. In her memoir, *The Tenth Muse*,
Jones recalls:

> *One day, when Frank had gone off into the heart of Paris
> for a literary lunch, I set to work on a pile of submissions that
> he wanted rejected. As I made my way through, I was drawn
> to the face on the cover of a book that Calmann Lévy was
> about to publish. It was the French edition of* Anne Frank:
> The Diary of a Young Girl. *I started reading it—and I
> couldn't stop. All afternoon, I remained curled up on the sofa,
> sharing Anne's life in the attic, until the last light was gone
> and I heard Frank's key at the front door. Surprised to find
> me still there, he was even more surprised to hear that it was
> Anne Frank who had kept me. But he was finally persuaded
> by my enthusiasm and let me get the book off to Doubleday in
> New York, urging them to publish it.*
>
> *It didn't take much urging, and we were given the go-
> ahead to offer a contract.*

Among those who, early on, recognized the book's importance was Robert Warshaw, an editor at *Commentary*, a highly
regarded Jewish-interest magazine that printed excerpts from
the diary in advance of its American publication. "Let me say
again," Warshaw wrote Otto Frank, "that I have read no document of the Jewish experience in Europe that seemed to me so

expressive, so moving, and on so high a literary level as your daughter's remarkable diary."

But despite the enthusiasm of Warshaw and other early readers, Doubleday's ambitions for the book were modest. The publisher agreed to pay Otto Frank a $500 advance, and a small print run was ordered

The book's American editor was a young woman named Barbara Zimmerman, later Barbara Epstein, later still a founder of the *New York Review of Books*, who was then around the age that Anne Frank would have been had she lived. Her correspondence with Otto Frank is a model of personal affection, professional savvy, and faith in the importance of the project on which they were collaborating. On November 5, 1951, Zimmerman wrote Otto Frank, "I love the book and feel that it has a value for me beyond matters of business."

Every decision concerning the packaging and the launch of *The Diary of a Young Girl*, which was published on what would have been Anne's twenty-third birthday, turned out to have been an inspired one. Good fortune and serendipity appeared, at every stage, to arrange Anne's diary's American success.

Some of the credit is Doubleday's, and some is Otto Frank's, who rather quickly caught on to the publishing business and to the business of publicity in particular. He realized that his daughter's diary was not in fact the relic of a saint, not the sanctified remains of what Ian Buruma, writing in the *New York Review of Books*, called the "Jewish Joan of Arc," but simply a book. Pained at first by the unpleasant side effects of commercialization, Otto learned to steel himself to the discomfort of having his daughter talked about as if she were a fictional character. As always, he was determined to support his family, which would soon include his second wife, Elfriede "Fritzi" Markovits Geiringer, whom he married in November 1953.

When it became clear to Otto that the diary was becoming not merely a commodity but a lucrative one, he decided to channel some of the profits it generated into the human-rights causes that would become, to him, as much of a religion as the Reform Judaism he practiced after the war. As soon as Otto saw what the diary could accomplish, he became quite single-minded—practical, focused, and at least partly immune to second thoughts or distractions.

IT MUST have been an obvious choice to put Anne's face on the cover, and Otto Frank sent his publishers a photograph of his photogenic daughter. Before the war, he had been a passionate amateur photographer. With his Leica camera, one of the first to be sold commercially, he documented births, birthdays, family holidays, and vacations, marking each stage of his daughters' development with dozens of formal portraits and snapshots of the girls brushing their teeth, combing their hair, playing with friends, sunbathing, building sand castles. Scores of photos survived the war, striking visual images that would contribute to Anne Frank's celebrity.

For the American edition, Otto selected a picture taken in 1939. In the photo, among the most sedate of Anne's portraits, her beautiful face conveys a wistful intelligence and a piercing sweetness. It was the picture that Anne had pasted in her diary, with a note remarking that such a portrait might improve her chances of getting into Hollywood. In real life, she added, she often looked quite different. She was probably correct, if we assume that the majority of her photos—in which she is shown laughing or smiling impishly, more animated and funny faced than conventionally pretty and composed—provide a more accurate likeness. But she preferred to be seen as a serious, lovely girl, and in choosing the picture that she herself picked, Otto may have felt that he was again fulfilling her wishes. In later versions of the diary, the current paperback edition, and other books about Anne, more cheerful images have been used.

In a review in the *New Statesman* in May 1952, Antonia White responded, as so many have, to the photo: "What she has left behind is a book of extraordinary human and historical interest, as living as the mischievous, intelligent face in the photograph which confronts the middle-aged reader with the same shrewd pertness that must so often have been turned on her parents and the Van Daans."

It's impossible to overestimate the power that Anne Frank's image has had. She is instantly identifiable, whether we see her face on a book or projected (to coincide with a visit of the Anne Frank traveling exhibit) on a tower in Great Britain where Jews were tortured during the Middle Ages. It is an understatement to say that she is the single most commonly recognized and easily recognizable victim of the Nazi campaign against the Jews, or of any genocide before or since. The passions that she has invoked cannot be separated from the fact that we know what she looked like.

Anne's author photo was a publisher's dream. At Doubleday, Donald B. Elder wrote to thank Otto for the "very charming" picture. Pleased by the portrait, Barbara Zimmerman must have felt her hope for the book take another quantum leap when she managed to secure a brief introductory essay from Eleanor Roosevelt.

Over fifty years later, this preface still introduces the book, even though Americans have since learned and forgotten that many Jews—including Otto Frank's family—failed to find refuge in the United States in part because of the policies of Mrs. Roosevelt's husband. Meanwhile, the teenage author's fame may have outdistanced that of her introducer; by now, it seems likely that more American schoolchildren have heard of Anne Frank than of Eleanor Roosevelt.

The essay of just over a page begins, "This is a remarkable book. Written by a young girl—and the young are not afraid of telling the truth—it is one of the wisest and most moving commentaries on war and its impact on human beings that I have ever read." The prologue stresses the triumph of the spirit that the diary documents and the ways in which its author's concerns resemble those of teenagers everywhere. "Despite the horror and humiliation of their daily lives, these people never gave up . . . Anne wrote and thought much of the time about things which very sensitive and talented adolescents without the threat of death will write—her relations with her parents, her developing self-awareness, the problems of growing up."

The words "Jew" or "Jewish" are never mentioned. It has been remarked that Eleanor had grown up in a milieu pervaded by what Roosevelt biographer Geoffrey C. Ward termed a "kind of jocular anti-Semitism." After a friend of the Roosevelts caught a large jewfish on a 1923 fishing expedition, Eleanor quipped, to her husband's amusement, "I thought we left New York to get *away* from the Jews."

Moved by the former first lady's preface, Otto Frank wrote Mrs. Roosevelt, thanking her for her kind words: "Reading your introduction gives me comfort and the conviction that Anne's wish is fulfilled: to live still after her death and to have done something for mankind."

Like so much else about Anne's diary, this preface has been the subject of controversy, in this case involving the charge that the book's American editor wrote the introduction *for* the former first lady and asked Mrs. Roosevelt to sign it. But if that were true, Barbara Zimmerman didn't say so to Otto Frank, to whom she conveyed her delight in Mrs. Roosevelt's foreword. And Otto's letter of thanks to Mrs. Roosevelt (who would encourage him to allow a stage or film version to be made from the diary, so that Anne's message could reach a wider audience) was purely sincere. He described his sense of "mission in publishing her ideas, as I felt that they help people to understand . . . that only love not hatred can build a better world." It's a touching correspondence, as is a later exchange in which Otto declines an invitation to meet the first lady during her stay at the Park Sheraton in Manhattan on the grounds that he has recently suffered a nervous breakdown and needs to take a little rest.

Less sweet was Mrs. Roosevelt's readiness to believe the charges in a letter she later received from a writer who accused Otto of, among other things, having moved to Switzerland to avoid paying high Dutch taxes. The writer of that letter was an American novelist named Meyer Levin, who had given the diary a rave review on the front page of the *New York Times Book Review*.

THE BOOK was an instant sensation. Meyer Levin's review sold it.

Not since Zelda Fitzgerald critiqued (pseudonymously and negatively) a book by her husband had there been a literary review assignment—given, accepted, or, in this case, requested—in which questions of conflict of interest so blatantly arise. Meyer Levin was not only a close friend of Otto Frank's, but he was acting as Otto's adviser, and, informally, as the diary's agent. In addition, he was convinced that he was the perfect choice to adapt the play for the stage.

Nevertheless, he asked permission to write the essay, and Francis Brown, the assigning editor, agreed, then later gave him more space for line after line of praise: "For little Anne Frank, spirited, moody, witty, self-doubting, succeeded in communicating in virtually perfect, or classic, form, the drama of puberty." While acknowledging the painful subject matter, Levin anticipated his readers' reservations, which he preemptively dispelled, assuring them that "this is no lugubrious ghetto tale, no compilation of horrors . . . Anne Frank's diary simply bubbles with amusement, love, discovery . . . These people might be living next door;

their within-the-family emotions, their tensions and satisfactions are those of human character and growth, anywhere."

The variant versions of the diary, including Anne's revisions, would not be available in English for another thirty years, and Levin perpetrated the most common myth—or partial truth—about Anne's work: "Because the diary was not written in retrospect, it contains the trembling life of every moment."

Other critics were equally enthusiastic. *Time* magazine called Anne's book "one of the most moving stories that anyone, anywhere, has managed to tell about World War II." On the same page as a review of Flannery O'Connor's first novel, *Wise Blood*, the Catholic journal *Commonweal* praised the diary as "extraordinary for its writer's candor and sensitivity, both to her environment and her interior development."

The official publication date was June 12, and on June 23, Barbara Zimmerman wrote Otto Frank that the first edition had sold out; a second and third printing of ten thousand copies each had been ordered. The house had decided to go all out on ads and promotion. She was certain that the book would be a huge best seller, and the warm public response had renewed her faith in the American people. "ANNE FRANK is a tremendous success . . . ," wrote Zimmerman. "It is one of the biggest books that has been published in America for a long while. Simply working on this book has been a most wonderful experience for me because I am quite frankly in love with it! And it is so nice to find so many hundreds of others who agree!"

Part of what makes Barbara Zimmerman's letters to Otto Frank so sympathetic and so touching is that they make it possible to imagine what it was like to be in your early twenties and get your first real publishing job in New York, and one of the first books you are assigned to edit happens to be *The Diary of a Young Girl*.

FOUR

The Book, Part II

UNLIKE THOSE BOOKS THAT WE LOVED AS CHILDREN
and return to as adults with the bewilderment of someone
visiting the site of a childhood home that has been torn down
to make room for a superhighway, Anne Frank's diary never
makes us wonder: Who was that person who liked this book?
Rather, like any classic—it may be one definition of a literary
classic—it rewards rereading. Each reading (I am referring here
to the "c" version of the diary, which Otto Frank assembled by
combining Anne's first draft and her revisions, the edition that
schoolchildren read and that most of us first encountered) reveals aspects of the work that we may have missed before and
allows us to view the book in the light of our own experience,
of everything we have learned, remembered, and forgotten
since the first time we read it.

Though most young readers might not know what to call
it, or how to identify the source of the book's appeal, the first
thing that draws us into the diary is Anne Frank's voice, that

mysterious amalgam of talent, instinct, hard work, and count-less small authorial decisions that make words seem to speak to us from the page. The assured, infectious energy of that voice makes us willing, even eager, to hear a little girl tell us what gifts she got for her thirteenth birthday and how her friends watched a Rin Tin Tin film at her party. We are patient, even charmed, as the child prattles on about who her best friend is now as opposed to which girl *used* to be her best friend, which boy she has a crush on, which boy she intends to marry.

One of the misconceptions about *The Diary of a Young Girl* is the notion that, from the beginning, Anne called her diary Kitty. In fact, in the early drafts, she framed some entries as let-ters to friends—some real, some imaginary—with whom she kept up a lively, if one-sided, correspondence. In one affecting note, Anne tells a friend that this will be the last letter she will be able to send. Other letters were addressed to characters in Cissy van Marxveldt's *Joop ter Heul* novels, a popular series of books, of which Anne was extremely fond.

The series, writes Mirjam Pressler, "follows the fortunes of a 'club' of girls from school to marriage to motherhood. The subjects of the books are not very different from those of the girls' books published elsewhere in the world at the same time—stories with an almost educational feel to them, preparing girls for their future roles as wives and mothers. In style, however, they are quite different—more colloquial and amusing; it is tempting to say more modern . . . We may safely assume that Cissy van Marxveldt had some influence on Anne's own style." In September 1942, during a period when Anne mentions read-ing the *Joop ter Heul* books, she addresses her diary letters (later cut or changed in her revisions) to Conny, Marianne, Phien, Emmy, Jettje, and Poppie—members of the "club."

One of Van Marxveldt's heroines was Kitty Francken, and it was Kitty on whom Anne decided when, during her last months in the attic, she began to revise her diary and focused on one imaginary listener. Though Anne had had a real friend by that name, Käthe "Kitty" Egyedi, it is generally agreed that Anne chose the name from among Van Marxveldt's characters. Anne may have envisioned *Het Achterhuis* as a *Joop ter Heul*–style romance of the secret annex.

What matters is that this device—the diary letters to Kitty—gave Anne a way of addressing her readers intimately and directly, in the second person: *you you you*. Perhaps it helped her write more fluently by providing her with an imaginary audience. Many people have found themselves prevented from keeping a diary or journal by uncertainty and confusion about whom exactly the diarist is supposed to be writing *to* or *for*.

Reading Anne's diary, we become the friend, the most intelligent, comprehending companion that anyone could hope to find. Chatty, humorous, familiar, Anne is writing to us, speaking from the heart to the ideal confidante, and we rise to the challenge and become that confidante. She turns us into the consummate *listener*, picking up the signals she hopes she is transmitting into the fresh air beyond the prison of the attic. If her diary is a message in a bottle, we are the ones who find it, glittering on the beach.

Within a few pages, the transparency of Anne's prose style has convinced us that she is telling the truth as she describes the world around her and looks inward, as if her private self is a foreign country whose geography and customs she is struggling to understand so that she can live there. Among the motifs that run throughout the book is Anne's urgent desire to find out who—what sort of person—she is.

The subject of Anne's true nature absorbs her, and us, from the earliest passages to the diary's final entry, in which she talks about her "dual personality," the lighthearted, superficial side that lies in wait to ambush and push away her "better, deeper, and purer" self. Aware of how often she hides her good qualities because she is afraid of being misunderstood or mocked, she accuses herself of being uncharitable, supercilious, and peevish. "I twist my heart round again, so that the bad is on the outside and the good is on the inside, and keep trying to find a way of becoming what I would so like to be, and what I could be, if . . . there weren't any other people living in the world." On the brink of a horror of which she was painfully aware, Anne Frank agonizes about not being nicer to her mother.

Reading her diary, we are reminded, often shockingly, of similar questions we may have asked ourselves as adolescents, of the enthralling mysteries that time has solved, or whose urgency age has erased. It's nearly impossible to recall wondering who we were, who we *really were*, as well as our related preoccupations with the differences between our authentic selves and the outer shells that everyone mistook for the real thing. Trying to remember the psychological and spiritual contortions we put ourselves through, when we were young, is as difficult as trying to summon back our astonishment at how quickly our bodies were changing.

Perhaps more than any other book, Anne's diary reminds us of what that bewilderment and yearning were like. Meanwhile, the diary entries become a sort of mirror in which teenagers, male and female, can see themselves—a capsule description of the alienation, the loneliness, and the torrents of free-floating grief that define adolescence in twentieth-century Western culture. Older readers will recognize familiar but forgotten echoes from their own pasts as Anne describes her inability to breach the wall that separates her from others. Younger readers may experience an almost eerie kinship with a girl who died so long ago but who is saying what no one has expressed quite so succinctly. Of course, she is writing about eight Jews forced by the

Nazis to spend two years in an attic. But she is also describing what it is like to be young.

Among the fascinations of the diary is the chance it offers to watch Anne's protagonist and narrator—herself—revealed in all her complexity, and to witness what John Berryman called the transformation of the child into the adult. Anne Frank was immensely observant, and unabashedly curious about everything from current events to the quirks of human nature to the problems of being a movie star to the sex life of a cat. She also had a highly developed sense of humor, which served her well during the worst moments in hiding. When Pfeffer arrives in the attic and brings news of the "gruesome and dreadful" fates of their Jewish friends and neighbors, Anne promises herself that "we shall still have our jokes and tease each other when these horrors have faded a bit in our minds; it won't do us any good, or help those outside to go on being as gloomy as we are at the moment. And what would be the object of making our 'Secret Annex' into a 'Secret Annex of Gloom'?"

The form of the diary—letters with breaks, like chapter breaks, allowing for gaps in time and changes of subject—lets Anne glide from meditation to action, from narration and reflection to dialogue and dramatized scene. Part of what keeps us reading with such rapt attention are the regular yet unpredictable shifts between opposites of tone and content—between domesticity and danger, between the private and the historic, between metaphysics and high comedy. One of the most intriguing of these oppositions is the tension between the extraordinary and the ordinary, the extreme and the normal, the young genius and the typical teen. In one entry, Anne can make the most trenchant or poetic observations; in the next, she complains that she is being picked on, singled out, criticized unfairly; the adults don't understand her, they treat her like the child that she sounds like in these passages. Even as the dangers grew more pressing and her reflections more transcendent, she keeps insisting on how ordinary she is, and regardless of the evidence to the contrary, we believe her, and we don't, because it's true and it isn't.

Her voice is so recognizable and so evocative that we might mistake it for any girl's, until we read more closely and realize that its timbre, its tempo, and its choice of what to focus on is uniquely Anne's. Anyone who has ever tried to write autobiographically will know how difficult it is to do so without seeming mannered, strained, and false. Only a natural writer could sound as if she is not writing so much as *thinking* on the page.

"I have one outstanding trait in my character, which must strike anyone who knows me for any length of time, and that is my self-knowledge. I can watch myself and my actions, just like an outsider." Anne's self-scrutiny occasionally leads her to write about herself in the third person, as if she is describing an out-of-body experience during which she is watching herself interact with the others or simply lie in bed. "Then a certain person lies awake for about a quarter of an hour, listening to the sounds of the night. Firstly, to whether there might be a burglar downstairs, then to the various beds, above, next door, and in my room, from which one is usually able to make out how the various members of the household are sleeping, or how they pass the night in wakefulness."

What makes these moments of detachment all the more affecting is that they are often associated with the desire to escape the semiconstant state of terror in which she and her family exist. Among the diary's most lyrical passages is one in which Anne, who has just been startled by a loud ring at the door, envisions the fragile perch on which she and the others are huddled:

I see the eight of us within our "Secret Annex" as if we were a little piece of blue heaven, surrounded by black, black rain clouds. The round, clearly defined spot where we stand is still safe, but the clouds gather more closely about us and the circle which separates us from the approaching dangers closes more and more tightly. Now we are so surrounded by danger and darkness that we bump against each other, as we search desperately for a means of escape. We all look down below, where people are fighting each other, we look above, where it is quiet and beautiful, and meanwhile we are cut off by the great dark mass.

Even as she gropes her way through that great dark mass, Anne is remarkably restrained in calibrating the amount of fear she will admit into the diary. The air raids, the break-ins, and the brutality reported by the helpers and glimpsed from the window appear at regular intervals, so that the reader can never fully relax. Anne's open-hearted compassion is so powerful and contagious that she makes us feel, as she does, for the elderly, crippled Jewish woman whom Miep has seen sitting on a doorstep, where the Gestapo ordered her to wait while they found a car to take her away.

Until late 1943, when Anne's fear and anxiety spike, she tends to underplay the gravity of her situation and often ends a disturbing section with a consolatory joke. She appears to be reassuring Kitty, and, at the same time, herself. Her optimism, such as it is, seems like the pure product of youth and inspires a tenderness that few readers feel on reading the war diaries kept by, among others, Mikhail Sebastian, Viktor Klemperer, and Etty Hillesum. Written wholly or partly while their authors were in the world—the final section of Hillesum's book is composed of letters from Westerbork—these brilliant eyewitness accounts involve numerous locations and large casts of characters, few of whom are as memorable as the Franks, the Van Daans, and Dussel. More comprehensive than Anne's, offering views of their times that tend more toward the panorama than the keyhole, these journals were written by complicated adults, and each book, for different reasons, is as easy to admire but harder to love than the one by the no less complicated child.

WHAT I could be, if . . . there weren't any other people living in the world. No one would have, or could have, planned it so that Anne Frank ended her book this way—no more than it could have been arranged that she received her diary as a birthday gift and almost instantly began to write in it, so that the book begins not in the dark confines of the attic, where the constrictions and deprivations are already making themselves felt, but rather in the bright light that in those years passed for normality. In the daylight we can see what kind of person Anne was—who she was before, and might have become without, the incarceration in the attic.

The first entry in Anne's diary (again, in the version her father edited) nearly lifts itself off the page, powered by the joy that a life-loving, theatrical girl feels at the dawn of her thirteenth birthday. She's practically bursting out of herself, awake at six in the morning to see her presents. But she must stay in bed until seven, when she is allowed to get up and unwrap her gifts: roses, a plant, some peonies, books, a puzzle, a brooch, money, sweets.

She lists her new books—*Tales and Legends of the Netherlands, Daisy's Mountain Holiday*—and another she intends to buy with her birthday money, *The Myths of Greece and Rome.* We can tell what kind of girl she is: a reader, a fan of legends and adventure stories, of fantasy and the imagination. Of course, the gift she mentions first is the "nicest" of all, the diary that will later be given the name that history remembers: *Kitty.*

Our trust in Anne as a narrator will prove increasingly important as she describes daily life in a deceptively gentle circle of hell that, without her as our Virgil, we could hardly imagine. In the early diary entries, intimations of dread and peril (conveyed by the catalog of quotidian things that Jews are forbidden) alternate with equally quotidian activities—Ping-Pong games, flirtations, classroom dramas—that Anne is still able to enjoy. Our image of her as an "incurable chatterbox" whom her exasperated teacher assigns to write a composition called "'Quack, quack, quack,' says Mrs. Natterbeak" will inform our sense of her character and increase our sympathy for her neighbors in the attic, who must bear up under the strain (which Anne would be the last to notice) of her irrepressible conversation. Often, in these first diary entries, the descriptions of simple pleasures and of the punitive, shaming Nazi regulations appear within a single paragraph, such as this chilling variation on the theme of a parent worried by a child's lateness:

"Harry visited us yesterday to meet my parents. I had bought ice cream cake, sweets, tea, and fancy biscuits, quite a spread, but neither Harry nor I felt like sitting stiffly side by side indefinitely, so we went for a walk, and it was already ten past eight when he brought me home. Daddy was very cross, and thought it was very wrong of me because it is dangerous for Jews to be out after eight o'clock, and I had to promise to be in by ten to eight in future."

On occasion the contradictions of trying to live normally under abnormal circumstances are compressed into a sentence: "We ping-pongers are very partial to ice cream, especially in summer, when one gets warm at the game, so we usually finish up with a visit to the nearest ice-cream shop, Delphi or Oasis, where Jews are allowed."

ANOTHER factor that contributes to the diary's power to move us and to make us remember so much of what Anne tells Kitty is Anne's eye for detail, for the gesture or line of dialogue that forms and refines her portraits of her family and neighbors so they become three-dimensional characters in a work of art.

In "The Development of Anne Frank," John Berryman offers an example of Anne's dispassionate observation, a passage in which she refers to her father by his nickname, Pim: "She was vivacious but intensely serious, devoted but playful . . . imaginative yet practical, passionate but ironic and cold-eyed. Most of the qualities that I am naming need no illustration for a reader of the *Diary*; perhaps 'cold-eyed' may have an exemplar: 'Pim, who was sitting on a chair in a beam of sunlight that shone through a window, kept being pushed from one side to the other. In addition, I think his rheumatism was bothering him, because he sat rather hunched up with a miserable look on his face . . . He looked exactly like some shriveled up old man from an old people's home.' So much for an image of the man—her adored father—whom she loves best in the world. She was self-absorbed but un-self-pitying, charitable but sarcastic, industrious but dreamy, brave but sensitive."

Anne's diary abounds in illuminating details—of setting, of action and repose, of food and clothing, of mood, of conversation and response. In case we have trouble visualizing the architecture of her hiding place, Anne maps it out for us and helps us understand where each room—each public space that will also serve as private quarters for working and sleeping—is located in relation to the others.

In an entry dated August 4, 1943, Anne begins an hour-by-hour account of what, after a little more than a year, has come to constitute an ordinary day in an existence that is "so different from ordinary times and ordinary people's lives." Every aspect of the daily routine in the annex is made use of for what it reveals about the quirks and personalities of the people forced to follow the intricate steps of the harrowingly restrictive choreography.

Anne starts her timetable at nine in the evening, when the cacophony of preparations for the night reaches a crescendo in the thunderous sounds of Mrs. Van Pels's bed being moved to the window, "in order to give Her Majesty in the pink bed jacket fresh air to tickle her dainty nostrils!"

Washing up in the bathroom, Anne notices a tiny flea floating in the water. When gunfire erupts in the darkness outside, she wakes, "so busy dreaming that I'm thinking about French irregular verbs" until she realizes what she is hearing and creeps, for comfort, to her father's bed.

At lunchtime, when the warehouse workers leave and the annex residents can briefly relax, Mrs. Van Pels pulls out the vacuum cleaner and tends her "beautiful, and only, carpet," while Otto retreats to a corner to escape into the novels of his beloved Dickens. Finally, the workday ends, the helpers come upstairs, a radio broadcast silences even the loquacious Mrs. Van Pels. After a nap, it's time to gather for dinner, a scene that Anne documents at length.

Two weeks later, in a "Continuation of the 'Secret Annex' daily timetable," Anne returns to the subject of her father's love for Dickens, and this time, uses this detail to convey something seemingly trivial—but in fact revealing—about her parents' marriage.

Otto keeps trying to interest his wife in what he has been reading, but she insists that she doesn't have time. As if there were anything *but* time in the secret annex! When he makes another attempt, she suddenly remembers something she needs to tell one of her daughters—and a potentially companionable moment between a husband and wife has ended in a standoff.

Anne's revealing focus on the minutiae of daily life reminds the reader of how cautious the attic residents had to be about trivial things, and of how the need for such vigilance must have sharpened Anne's eye. "Although it is fairly warm, we have to light our fires every other day in order to burn vegetable peelings and refuse. We can't put anything in the garbage pails, because we must always think of the warehouse boy. How easily one could be betrayed by being a little careless!"

Days earler, Anne had trained her attentive gaze on the decline in the standard of living—their "manners," she calls them—in the annex. The oilcloth they have been using continually on the communal table has grown dirty. The Van Pelses have been sleeping all winter on the same flannelette sheet. Otto's trousers are frayed, and his tie is worn. Edith's corset has split and can no longer be repaired, and Margot is wearing a brassiere two sizes too small.

The following January, Anne entrusts Kitty with this inspired and withering complaint about how tired she has grown of the grown-ups' conversation—a seemingly lighthearted account that captures the stultifying tedium of social life in a place whose residents can no longer find anything new to say: "If the conversation at mealtimes isn't over politics or a delicious meal, then Mummy or Mrs. v.P. trot out one of the old stories of their youth, which we've heard so many times before, or Pf. twaddles on about his wife's extensive wardrobe, beautiful race horses, leaking rowboats, boys who can swim at the age of 4, muscular pains and nervous patients. What it all boils down to is this, that if one of the eight of us opens his mouth, the other seven can finish the story for him! We all know the point of every joke from the start, and the storyteller is alone in laughing at his witticisms. The various milkmen, grocers and butchers of the two ex-housewives have already grown beards in our eyes, so often have they been praised to the skies or pulled to pieces; it is impossible for anything in the conversation here to be fresh or new."

Anne defines the people around her by noting their different solutions to a problem, or their diverse answers to a single question. Early in the diary, the arrangements for bathing— an activity that every attic resident approaches differently— provide a series of clues to their personalities, and to the extent to which they have adjusted to their new lives. Peter, Anne tells us, chooses to bathe in the kitchen even though it has a glass door and he is so modest that, before each bath, he goes around to each of the annex residents and warns them not to walk past the kitchen for half an hour. Mr. Van Pels cherishes his privacy enough to carry hot water all the way upstairs. Uncertain about how best to carry out this delicate and newly demanding activity, Mrs. Van Pels has avoided bathing at all until she figures out the most convenient and comfortable place. Otto washes up in his private office, Edith behind a fire guard in the kitchen, while Margot and Anne retreat to the front office. Peter has suggested that Anne use the large bathroom in the office, where she can turn on the light, lock the door, and be alone. "I tried my beautiful bathroom on Sunday for the first time and although it sounds mad, I think it is the best place of all."

At dinner, during her "daily timetable" of life in the attic, Anne goes around the table, differentiating her characters by telling us what and how each person eats. Mr. Van Pels generously helps himself first, meanwhile offering his "irrevocable" opinion on every subject. His wife picks over the food, taking the tiniest potatoes, the daintiest morsels, smiling coquettishly and assuming everyone is interested in what she has to say. Their son eats a great deal and hardly speaks. Margot is also silent, though she "eats like a little mouse." Mummy: "good appetite, very talkative." Otto makes sure that everyone is served before he is, and that the children have the choicest portions.

Pfeffer ("helps himself, never looks up, eats and doesn't talk") provokes, from Anne, a diatribe that progresses from the "enormous helpings" he takes to his habit of hogging the bathroom when others need to use it.

A year into their stay in the annex, the residents play a game. If they were free, what would they do first? Like any author who has learned that an effective way to create a character is to indicate that person's hopes and fears, Anne reports each person's fantasies of liberation. Margot and Mr. Van Pels dream of a hot bath, at least a half hour long. Mrs. Frank longs for real coffee. Mrs. Van Pels wants ice-cream cakes. Peter longs to go to town and to the movies. Anne wants a home, the ability to move around freely, and to have some help with her work, by which she means school; this last is a somewhat odd and perhaps even thoughtless wish, since we know that Otto has been supervising the girls' lessons and, in theory, giving them all the help they need

When Otto's turn comes, he says he would choose to visit Mr. Vossen, Anne's pseudonym for Bep's father, Mr. Voskuijl. A month before, the residents had learned that Johannes Voskuijl, who had built the bookcase that camouflaged the entrance to the annex, had been diagnosed with stomach cancer and was not expected to recover. He must have been very much on everyone's mind. We also know that on the afternoon Margot received her call-up notice, Otto was visiting the Jewish hospital for the indigent elderly. Comforting the poor and old was something Otto did; it was among the reasons he was so admired. But when, in the game, he makes that choice, the reader can imagine how, at moments, it might have been a trial for his wife and daughters to live with this pillar of moral perfection. Couldn't he have picked the hot bath and *then* the hospital visit?

The final survey of this sort occurs in March 1944, when Anne polls her neighbors for their responses to their depressingly deficient diet. By now, we know her characters so well that we can almost predict their replies. Mrs. Van Pels complains bitterly about the difficulty of cooking with limited ingredients, and about the ingratitude she receives in return for all her hard work. Her husband claims he can stand the bad food as long as he has enough cigarettes. Edith replies that food is not so important to her, but she would love a slice of rye bread, and, incidentally, she thinks that Mrs. Van Pels should put a stop to her husband's smoking. Otto says that he not only needs nothing, but that part of his ration should be saved for Elli. And Pfeffer's maddening bluster trails off in ellipses. . . .

If we try to understand why we come to know these people so well, one explanation can be found in the patient accretion of actions and gestures with which Anne informs our vision of them. You can track each character through the book, watching their portraits emerge like photos coming up in a tray of developing fluid. It often seems as if Anne is conscious of who she has been including or ignoring, of who has temporarily captured or lost her attention. Almost as soon as we become aware that one of the attic residents has fallen silent, or has briefly gone unnoticed, that character is brought in, center stage, to reassert the oppressive reality of his or her constant presence.

Unsurprisingly, given Anne's age, her parents are the object of almost as much intense scrutiny as she devotes to exploring the mystery of her essential self. She consistently uses one parent to define the other: Daddy is kind and patient, Mummy short-tempered and sarcastic; Daddy is transparent and sensitive, Mummy opaque and obtuse.

A diary kept by Edith Frank or Auguste van Pels might have painted a slightly different picture of Otto Frank, but Anne's perspective is the only one we have. In her view, Otto—"Pim"—is invariably dignified and fair, defending his daughters when they are being maligned, yet perfectly impartial when he must mediate a dispute. He is the educator, the peacemaker, the leader to whom the others bring their dissatisfactions, fears, and complaints. In a passage that Otto cut, we see him unclogging the communal toilet. When there is a burglary, Otto and Peter are the ones who go downstairs to investigate. "We must behave like soldiers," he tells the frightened Mrs. Van Pels.

Urging Anne to be nicer to her mother, Otto appears to take no satisfaction in being the more popular—indeed, the adored—parent. Anne worships her father, and in our own more jaded and suspicious era, her diary serves as a useful reminder of how an adolescent daughter can feel passionately about her father without their relationship bordering on the incestuous or improper. This too may be one reason the diary has remained popular among young readers—its honesty about emotions that teenagers have learned to keep private.

Before Anne's romance with Peter takes its course and she tires of him, there is a dramatic incident that begins when her father asks her not to spend evenings alone with Peter in his room. Otto, we may feel, is right to worry. His daughter is a precocious adolescent, Peter is several years older. A pregnancy would be disastrous. Enraged by what she interprets as her father's lack of faith in her, Anne decides what she wants to tell him. She writes a note saying that she has reached a stage at which she can live entirely on her own. Her father can no longer talk her out of going upstairs. Either he forbids her to be alone with Peter, or else he trusts her completely—and leaves her in peace. Then she slips the letter into Otto's pocket.

Otto replies that he has received many letters in his life, but this is the most unpleasant. The remainder of his response, which Anne reports in direct dialogue, is such a model of forbearance and understanding, so thoroughly infused with a guilt-inducing sense of injury (how could Anne mistreat the loving parents who have done nothing but help and defend her?) that she caves in from remorse, just as she was meant to. "This is certainly the worst thing I have ever done in my life . . . to accuse Pim, who has done and still does do everything for me—no, that was too low for words . . . And the way Daddy has forgiven me makes me feel more than ever ashamed of myself."

Less than a week later, the tension has dissipated, and we see the attic residents celebrating Otto's birthday, for which he receives, among other gifts, a book on nature and a biography of Linnaeus. Anne makes it clear that her love of literature is part of what she shares with her father, who suggests that she and Margot list all the books they read in hiding. In Anne's daily schedule, time was allotted for reading, and on Saturdays, the Dutch helpers brought more books, which the attic residents eagerly anticipated. In addition to books about history and geography, biographies, a five-volume history of art, a children's Bible, compendiums of mythology, and what we would now call "young-adult novels," Anne mentions works by Oscar Wilde, Thackeray, the Brothers Grimm, and Alphonse Daudet.

Anne conveys her mother's character, as she does her father's, primarily through dialogue and action supplemented by commentary. One of Anne's earliest mentions of Edith occurs in an entry dated October 29, 1942, as Anne describes literary gifts from both parents. Otto has given her the plays of Goethe and Schiller, from which he plans to read to her every evening, starting with *Don Carlos*. "Following Daddy's good example"— note the pointed irony of that phrase, which underlines the passage's significance—"Mummy has pressed her prayer book into my hand. For decency's sake I read some of the prayers in German; they are certainly beautiful but they don't convey much to me. Why does she force me to be pious, just to oblige her?"

Regardless of the degree to which Otto's editing modulated

Anne's criticisms, her estrangement from her mother is a constant theme, and is reflected in the novel on which she was at work in early 1944, *Cady's Life*, a portion of which appears in *Tales from the Secret Annex*. The book begins when Cady, who has been hit by a car, complains to a friendly nurse about her mother's tactlessness. Anne uses the license of fiction to be even harsher about a troubled mother-daughter relationship than she is in the diary.

"She talks so unfeelingly about the most sensitive subjects," complains Cady. "She understands nothing of what's going on inside me, and yet she's always saying she's so interested in adolescents. . . . She may be a woman, but she's not a real mother!" In response, the wise Nurse Ank (much like the "nice Anne" who Anne claims to keep hidden) replies, "Perhaps she's different because she's been through a lot and now prefers to avoid anything that might be painful."

A late diary entry (which Otto omitted from the edited version) includes an outline for the ending of *Cady's Life*. This summary follows the well-known passage in which Anne mentions her desire to become a journalist and her plans to publish *Het Achterhuis*. In the narrative Anne sketches, Cady marries a "well-to-do farmer" though she remains infatuated with her former sweetheart, Hans, whom she initially broke up with because he sympathized with the Nazis. The entry concludes, "It isn't sentimental nonsense for it's modeled on the story of Daddy's life"—a line that has been taken to mean that Anne knew, or at least believed, that the love of Otto's life was a woman he had known before Edith, and that his marriage to Edith had had more to do with her convenience than with passion.

When Anne tries to find the source of her antipathy to her mother, she dredges up a memory of Edith forbidding her to come along on a shopping expedition with Margot. Anne also refers to the maddening maternal sermons that remind her of how different she and her mother are. But unlike Mrs. Van Pels, whose irritating habits and character traits are documented by the many annoying things she says and does, "Mama Frank, champion of youth" behaves quite admirably at almost every juncture. We observe Edith defending her daughters from Pfeffer and the Van Pelses, keeping the peace, making sure that her children eat well, and so forth.

Part of what makes the diary feel so authentic is that, despite all her resolutions to improve her character, Anne makes only the most pro forma teenage effort to be fair and impartial about her mother. Edith gets no credit when she insists (over Otto's objections) that a candle be lit to comfort Anne, who has been frightened by the rattle of machine-gun fire. "When he complained, her answer was firm: 'After all, Anne's not exactly a veteran soldier,' and that was the end of it." Nor is Edith pitied when Anne's coldness makes her cry. Anne blames her mother for the distance between them, a gap that has been widened by Edith's thoughtless comments and tactless jokes, presumably at her daughter's expense. "Just as I shrink at her hard words, so did her heart when she realized that the love between us was gone. She cried half the night and hardly slept at all."

Miep Gies observed that Edith Frank often seemed depressed and withdrawn; when Miep left the annex, Edith would follow her downstairs and just stand there, waiting. Eventually, Miep realized that Edith wanted to talk to her in private. Unlike the others, who enjoyed discussing what they planned to do when the war ended, Edith was afraid that the war would never end.

In the diary, Edith can do nothing to mollify Anne, whose contempt for her mother has as much to do with who her mother *is* as with anything she *does*. Anne is appalled by the thought of growing up with the limited horizons, ambitions, and expectations of the women around her, and laments her own inability to respect her mother or to see her as a role model. Anne writes that she hopes to spend a year in Paris or London, studying languages and art history—an ambition she compares, with barely veiled contempt, to Margot's desire to go to Palestine and become a midwife.

In an essay entitled "Reading Anne Frank as a Woman," a feminist interpretation of Anne as "a woman who was censored by male editors," Berteke Waaldijk, a professor of Women's Studies at the University of Utrecht, points out a long and almost entirely overlooked passage that Otto Frank excised from the final section of the diary. Perhaps Otto assumed that a lengthy disquisition on women's rights might distract the reader heading into the final pages in which Anne is unknowingly hurtling toward her doom. At a point during which Anne was simultaneously writing new material and rapidly revising, she devoted a remarkable amount of space to the question of why women are treated as inferior to men:

"Presumably man, thanks to his greater physical strength, achieved dominance over women from the very start; man, who earns the money, who begets children, who may do what he wants . . . It is stupid enough of women to have borne it all in silence for such a long time, since the more centuries this arrangement lasts, the more deeply rooted it becomes. Luckily schooling, work and progress have opened women's eyes. In many countries . . . modern women demand the right of complete independence!"

All along, we have sensed that Anne's rage at her mother and at Mrs. Van Pels has involved their inability to be—or even to seem—as brave and sensible and competent as the men. Only late in the diary does Anne understand that there is an actual person behind the abstract symbol of female limitation and servitude that she has so despised. Her digression about the problems of, and the disrespect for, women ends by suggesting that those who have gone through childbirth are entitled to gratitude and sympathy on that score alone. Anne seems to have realized that her mother is not entirely to blame for the ways in which she had been conditioned to behave, and for her stunted ambitions and expectations.

Almost a year and a half after writing a particularly furious passage in which she describes forcing herself to remain calm and having to suppress the desire to slap her mother, Anne had second thoughts about including it in *Het Achterhuis*. In the meantime, she had undergone the change that John Berryman considered the turning point in her child-into-adult conversion, which immediately followed her vision of her friend Lies and her dead grandmother. "I was very unhappy again last evening. Granny and Lies came into my mind. Granny, oh darling Granny, how little we understood of what she suffered, or how sweet she was . . . And Lies, is she still alive? What is she doing? Oh, God, protect her and bring her back to us. Lies, I see in you all the time what my lot might have been."

In an entry dated three days later, January 2, 1944, Anne remarks that she has been rereading her diary and is shocked by the "hothead" sections about her mother. She blames her bitterness on "moods" that prevented her from seeing a situation from another person's point of view and from realizing that she might have hurt Mummy or made her unhappy. Her mother had responded in kind, and the result was "unpleasantness and misery rebounding all the time."

In the aftermath of her vision of Lies, Anne vows to improve. She promises herself that she will stop making her mother cry. She herself has grown more mature, and her mother is no longer quite so anxious. But a few days later, Anne is unable to refrain from returning to the theme of how hard it is to respect her mother and how little she wishes to follow her example.

The winter of 1944 marked the start of an acutely introspective period during which Anne looked back and measured the person she had become against the girl she was: "If I think of my life in 1942, it all feels so unreal. It was quite a different Anne who enjoyed that heavenly existence from the Anne who has grown wise within these walls. Yes, it was a heavenly life. Boyfriends at every turn, about 20 friends and acquaintances of my own age, the darling of nearly all the teachers, spoiled from top to toe by Mummy and Daddy, lots of sweets, enough pocket money, what more could one want? . . . I look now at that Anne Frank as an amusing, jokey, but superficial girl who has nothing to do with the Anne of today."

Had Anne survived, or had she been able to stay in contact with the wider world, she might have taken consolation from the discovery that many, if not most, teenage girls come into conflict with their mothers. Isolated in the attic, Anne could only examine her own history and her own conscience, and try to locate the wellspring of her sadness and her rage.

JUST as Anne finds a shorthand in which to express her complex relations with her parents by recording her response to their tastes in reading—Goethe versus the prayer book—so she introduces the Van Pelses by describing their arrival at the secret annex with Mrs. Van Pels's pottie in a hatbox, and her husband carrying a folding tea table under his arm. The Van Pelses begin their new lives with a noisy quarrel of the sort that Otto and Edith would never have had, and would certainly never have allowed to be overheard. Coarse, selfish about trivial matters, unembarrassed to squabble over plates and sheets, yet ultimately sympathetic, the Van Pelses do one thing after another that arouses, in the reader, amusement and affection commingled with annoyance. They are *characters*.

Because the Van Pelses are so much more transparent than the Franks, we can more easily watch them weakening and falling apart. By the last summer in the attic, Mrs. Van Pels is talking about hanging, suicide, prison, a bullet in the head. "She quarrels, uses abusive language, cries, pities herself, laughs, and then starts a fresh quarrel again."

Anne's response to Mrs. Van Pels is quite different from the helpers' memories of her real-life model, whom Miep Gies called, "a very uncomplicated person, anxious and cheerful at the same time." Miep came to see Mrs. Van Pels as not only realistic but prescient. "If anyone had a premonition of how badly it would all end, she was the one." Anne seems not to have known that, for Miep's birthday in February 1944, Auguste van Pels—who, in the diary, we see greedily holding on to every possession—gave Miep an antique diamond-and-onyx ring as a "way to express the inexpressible."

Mr. Van Pels grows more fractious as his cigarette supply dwindles. He is careless in ways that endanger the Jews and their helpers. Yet another crisis, marked by yet another fight, erupts when the Van Pelses' money runs out and they must sell Mrs. Van Pels's fur coat. It is a tribute to the vividness of Anne's writing that readers can recall the drama surrounding the loss of the coat decades after reading the diary. All over Europe, families were deciding what to sell or keep or barter as they struggled to survive. But of all those painful conversations, the one we hear about in detail is the one that Anne describes in a few lines. The argument erupts over a rabbit-skin coat that Mrs. Van Pels has worn for seventeen years, and for which her husband has received the impressive sum of 325 florins. Having hoped to save the money in order to purchase new clothes after the war, Mrs. Van Pels is enraged by her husband's insistence that the money is deperately needed by the household.

"The yells and screams, stamping and abuse—you can't possibly imagine it! It was frightening. My family stood at the bottom of the stairs, holding their breath, ready if necessary to drag them apart!"

How humiliating for the Van Pelses to have such a squalid fight with another family listening, and not just any family but the *perfect* Franks. In many ways, the Van Pelses are the more well drawn and rounded of the two couples in the secret annex, since—unlike the angelic Pim and (in Anne's view) the unfeeling Mummy—the Van Pelses are alternately and sometimes simultaneously maddening and touching.

The gunfire that frightens Anne terrifies her neighbor: "Mrs. Van Daan, the fatalist, was nearly crying, and said in a very timid little voice, 'Oh, it is so unpleasant! Oh, they are shooting so hard,' by which she really means 'I'm so frightened.'" There's something affecting about her husband's hypochondria, the "tremendous fuss" he makes about a little cold, rubbing himself with eucalyptus and gargling with chamomile tea. Anne's dual portrait captures so much that, even as enforced intimacy enrages her, we can see the Van Pelses' charm and vulnerability shining through.

One thing seems inarguable: Anne was able to make the Van Pelses so real and present to us that we grieve at the thought of the hand injury that made Hermann lose his will to survive at Auschwitz, just as we can hardly bear to wonder if Auguste regretted the loss of her beloved fur coat during that freezing march from Bergen-Belsen to her death.

IF the characterization of the Van Pelses is a marvel of literary portraiture, the image of their son, Peter, is another matter. If Peter strikes us as an interesting character, a closer reading reveals that this is largely because he is lit by the refracted glow of Anne's interest. When her fascination wanes and disappears, as it does in Anne's revisions—which we'll look at in the next chapter—we are left with only what we actually see him do and say. He accompanies Otto to investigate the break-in, does the heavy lifting of the sacks of beans, and wishes he weren't Jewish. Young readers may develop a crush on Peter, but it is Anne's crush. Her attraction transforms Peter into a romantic figure. But without that intensity—which, again, is Anne's—Peter is a touching but rather ordinary boy. Moody, mercurial, restless, not especially perceptive, he is a scrim on which the isolated girl can project her loneliness and longing.

Anne's early opinion of Peter is so harsh that one pleasure of reading the book is watching that antipathy reverse itself. In the "a" version of the diary, Anne reports getting a chocolate bar from Peter for her thirteenth birthday, before the families went into hiding. But the fact that they are acquaintances is hard to extract from her account of the Van Pelses' arrival in the attic; Anne calls Peter "a rather soft, shy, gawky youth; can't expect much from his company." The authors of the play must have thought it simplified matters to have the young couple meet for the first time in the annex, which is the impression that most readers and audience members come away with.

Over the next months, Anne emphasizes how boring, lazy, and hypersensitive Peter is; the hypochondria he shares with his father is less winning in a young person. The first dramatized scene in which he appears involves a fight over a book that his father doesn't want him reading. Peter gets credit for standing up to Mr. Van Pels, but loses it for the peevish and pouting quality of his resistance. Peter, we learn, has trouble with English, and has a comical fondness for using foreign words he doesn't understand.

By late September, Anne is telling the Van Pelses that Peter often strokes her cheek, and she wishes he wouldn't. Appalled by their response—could she grow to "like" Peter? He "certainly liked me very much"—Anne tells his parents that she thinks Peter is "rather awkward." But slowly a camaraderie develops; Peter and Anne both enjoy dressing up in the clothes of the opposite-sex parent. For his birthday, in November, Peter gets a razor, a Monopoly game, and a cigarette lighter—in contrast to the Franks, who get, and give one another, books. In fact, Anne tells us, Peter "seldom reads."

Peter at last moves to center stage as the hero of an adventure involving the transport of masses of beans. Four months later, we see him bitten by one of the large rats swarming the attic, and not long after, he is the one who goes downstairs with Otto after they hear a noise.

Not until January 1944 do we realize—before Anne does—what is starting to happen between the two teenagers: "It gave me a queer feeling each time I looked into his deep blue eyes, and he sat there with that mysterious laugh playing round his lips . . . and with my whole heart I almost beseeched him: oh, tell me, what is going on inside you, oh, can't you look beyond this ridiculous chatter?" By the next month, Anne and Peter are having the intimate conversations that will fuel Anne's longing for someone to love, as well as her conviction that this someone is Peter.

But even as Anne finds these exchanges endlessly fascinating, the reader may feel that Peter's contributions to these talks are less riveting than hers. Peter expresses his desire to go to the Dutch East Indies and live on a plantation, as well as his hope that he may be able to pass for Christian after the war. When he reveals his inferiority complex and claims he feels he is less intelligent than the Franks, an impartial observer might agree. And yet, early in March, Anne records wanting to do something about Peter's loneliness and sense of being unloved, and in a postscript to the March 6 entry, she admits that she has begun to live from one of their meetings to the next.

They talk, in the abstract, about kissing; they discover how much they have in common, how much they have changed during their time in the attic, how their ideas about each other have evolved. They discuss the fact that neither of them can confide in their parents, that frustration drives Anne to cry herself to sleep at night while Peter retreats to his loft and swears. They consider how different they were when they first arrived in the annex, and how they can barely recognize themselves as the same people they were in 1942. They marvel at the astonishing fact that they could have disliked each other at first, that Peter thought Anne chattered too much, while she was annoyed that he didn't bother flirting with her. When Peter refers to his tendency to isolate himself from the others, Anne tells him that his silence is, in a way, like her chatter. As unlikely as it may seem, she too loves peace and quiet. They admit how glad they are to be together, to have each other. And Anne tells Peter that she would love to be able to help him.

"'You always do help me,'" he said. 'How?' I asked, very surprised. 'By your cheerfulness.' That was certainly the loveliest thing he said. It was wonderful, he must have grown to love me as a friend, and that is enough for the time being . . .

"If he looks at me with those eyes that laugh and wink, then it's just as if a little light goes on inside me."

Such outpourings will be familiar (perhaps all too familiar) to anyone who has ever fallen in love. But they are utterly new to Anne, and, again, it is a tribute to her ability to write honestly and persuasively and to find the right tone for what she is telling Kitty (and us) that she can make it seem new. Anne longs for a kiss, they don't kiss, they kiss. How easy it would be for another writer to make this sound banal.

On May 19, Anne writes, "After my laborious conquest, I've

got the situation a bit more in hand now, but I don't think my love has cooled off." On June 14, she tells Kitty, "Peter is good and he's a darling, but still there's no denying that there's a lot about him that disappoints me." Three weeks later, Peter jokes about the possibility of becoming a criminal or a gambler, and Anne fears that Peter is becoming too dependent on her. "Poor boy, he's never known what it feels like to make other people happy, and I can't teach him that either . . . it hurts me every time I see how deserted, how scornful, and how poor he really is."

By the fifteenth of July, Anne's enchantment with Peter has reached a low ebb: "Now he clings to me, and for the time being, I don't see any way of shaking him off and putting him on his own feet. When I realized that he could not be a friend for my understanding, I thought I would at least try to lift him up out of his narrowmindedness and make him do something with his youth." That is the final mention of Peter in Anne's book.

In the theatrical and film versions of the diary, Anne and Peter are in the garret, staring rapturously at the heavens when the Gestapo come to arrest them. But that was not what happened. Anne was with her mother and sister. Otto was upstairs with Peter, helping him with the English lessons that, we know from Anne, gave him so much trouble.

Two relatively minor characters, the dentist Pfeffer and Margot Frank, are among the most nuanced and well drawn. With every minutely monitored tic, Fritz Pfeffer becomes the remarkable literary creation that is Albert Dussel. Of the eight people in the annex, his characterization is probably least like his counterpart in life; by all accounts, Pfeffer was extremely attractive to the ladies. But his charms were lost on Anne.

The dentist arrives late and brings bad news. Nazis have been going door-to-door, hunting down Jews. Friends have been rounded up and deported, loaded trucks rumble past, and columns of bullied prisoners trudge through the streets. These sobering truths, mixed with gratitude for having been spared the fate of her fellow Jews, temper the reluctance Anne otherwise might have felt on learning that Pfeffer—who is her father's age, but whom, unlike Otto, she refers to as *old*—is going to share her little room.

In her excellent biography of Anne Frank, Melissa Müller writes, "Otto and Edith's decision to put Pfeffer in the same room with Anne instead of with the sixteen-year-old Peter van Pels corroborates Anne's complaint that she was in fact regarded as a child. Not only Otto but Edith Frank as well disregarded her growing need for privacy and obviously ignored their adolescent daughter's sense of modesty, which was of course becoming all the more acute as she matured sexually."

Perhaps Anne's characterization of Pfeffer might have been a bit more sympathetic had she not spent night after wakeful night listening to a middle-aged man sleep. By contrast, Miep Gies very much liked her dentist, as did many of his loyal patients.

His Christian fiancée, Charlotte Kaletta, was devoted to Pffefer, who was nineteen years her senior and who had a son from an earlier marriage that had ended in divorce. The couple had lived in Germany until Hitler's racial laws forced them to flee in the futile hope that they could be married in Holland. In the diary, "Lotje" is referred to as Dussel's wife, and in a section cut from the "a" version, Anne mentions getting a roll of candy drops for her thirteenth birthday from "Mrs. Pfeffer." When her fiancé went into hiding, Charlotte kept up their correspondence, love letters that Miep delivered without revealing where he was.

After the war, Charlotte's friendship with Otto Frank ended, possibly because she was upset by Anne's portrayal of

Pfeffer in the diary and later by his characterization in the play. It's easy to see that the woman who loved Fritz Pfeffer so much that she waited for his return even after it had become clear that he had died in Neuengamme might object to the reader catching a near final glimpse of Pfeffer after a quarrel with the Franks over the "sharing out of the butter. Dussel's capitulation. Mrs. Van Daan and the latter very thick, flirtations, kisses and friendly little laughs. Dussel is beginning to get longings for women."

Anne's patience wears especially thin on Sundays, when Pfeffer performs the exercises she describes in appalled detail: "When he has ended with a couple of violent arm-waving exercises to loosen his muscles, His Lordship begins his toilet." Though everyone behaves as if no decent person should think twice about a grown man sharing a room with a pubescent girl, sexual discomfort suffuses Anne's view of Pfeffer. She's repulsed when she comes down with the flu and he plays doctor, laying his greasy head on her naked chest. "Not only did his hair tickle unbearably, but I was embarrassed in spite of the fact that he once, thirty years ago, studied medicine and has the title of Doctor. Why should the fellow come and lie on my heart? He's not my lover, after all! For that matter, he wouldn't hear whether it's healthy or unhealthy inside me anyway, his ears need syringing first as he's becoming alarmingly hard of hearing."

In July 1943, open warfare breaks out between the roommates when Pfeffer rejects Anne's "reasonable request" to use the little table in their room so she can work there, twice a week, from four until five thirty. He mocks her whole idea of work (mythology! knitting!). She asks Otto for advice, and she and the dentist attempt a détente. Pfeffer responds by berating Anne for her selfishness and her stubborn insistence on getting what she wants. Only when Otto intercedes do they agree: Anne can work in their shared room, two afternoons each week, but only until five. "Dussel looked down his nose very much, didn't speak to me for two days and still had to go and sit at the table from five till half past—frightfully childish."

Regardless of whether he needs it or not, Pfeffer insists on having all his allotted time at the contested table. Rarely in literature have we seen a more pointed illustration of human *smallness*, and of the inability to compromise with grace.

Anne is not the only person whom Pfeffer is driving mad. One evening, as the annex residents violate the unofficial prohibition against Teutonic culture and listen to a radio broadcast of "Immortal Music of the German Masters," the dentist fiddles with the dials until Peter explodes and Pfeffer replies, in his "most hoity-toity manner," that he is working to get the sound perfectly right. Yet Anne lets us see another side of Pfeffer when she records each resident's wish for what freedom will bring: "Dussel thinks of nothing but seeing Lotje, his wife."

In the diary, Pfeffer is given quite a lot to say, not nearly so much as Mrs. Van Pels, but far more than the good-girl Margot, who only rarely appears onstage, and who Anne interprets *for* us, interceding and telling us what her sister is like. Much of what we learn about Margot is the result of projection on Anne's part, as she repeatedly tries to intuit her sister's responses to life in the attic.

After her romance with Peter begins, Anne worries that her sister may also have feelings for the annex's only viable young male. Anne says, "I think it's so rotten that you should be the odd one out," to which her sister replies, "somewhat bitterly," that being the odd one out is something she's gotten used to. What does that mean? Anne doesn't ask, and either Margot doesn't say or Anne doesn't tell us.

Anne's first direct analysis of her sister comes at a moment when her sister has been held up (yet again, according to Anne) as a model human being. "I don't want to be in the least like Margot. She is much too soft and passive for my liking, and allows everyone to talk her around, and gives in about everything. I want to be a stronger character!"

More than a year afterward, Margot's "mouse-like" eating habits come up for scrutiny along with those of the others gathered around the table. "The only things that go down are vegetables and fruit. 'Spoiled' is the Van Daans' judgment; 'not enough fresh air and games' our opinion."

Later in the diary, Anne wisely allows Margot to speak for herself and reveal a facet of her character quite unlike the near saintliness with which she is often credited. Anne includes a letter in which Margot continues a discussion that she and Anne have been having about the possibility that Margot is jealous of Anne's involvement with Peter. Anne introduces the letter as "evidence of Margot's goodness," as if she were unaware of the barely veiled insult to Anne's intelligence that Margot can't help slipping into her explanation of why she could never feel close to Peter:

"I would want to have the feeling that he understood me through and through without my having to say much. But for that reason it would have to be someone whom I felt was my superior intellectually, and that is not the case with Peter. But I can imagine it being so with you and Peter."

"Not quite happy" with her sister's letter, Anne's answer includes a sentence that seems designed to elicit a flicker of sexual jealousy regardless of Margot's claim to feel none. "At present there is no question of such confidence as you have in mind between Peter and myself, but in the twilight beside an open window you can say more to each other than in brilliant sunshine."

In all the talk about Anne's symbolic and historical import, her spiritual development, her friction with her mother, her discovery of first love, little mention is made of how much her diary tells us about what it is like to have a sibling.

In an entry dated November 7, 1942, Anne describes the sort of family fight that has erupted in every household in which there is more than one child. Margot leaves her book around, Anne picks it up, Margot demands the book back, Anne wants it, her parents take Margot's side. This ensuing scene, which may be the closest we come to fully believing Anne's claim to ordinariness, is so familiar that we hardly notice how rarely it's done, or done well, in literature.

"Just because I wanted to look a little further on, Margot got more and more angry. Then Mummy joined in: 'Give the book to Margot, she was reading it,' she said. Daddy came into the room. He didn't even know what it was all about, but saw the injured look on Margot's face and promptly dropped on me: 'I'd like to see what you'd say if Margot ever started looking at one of your books!' I gave way at once, laid the book down, and left the room—offended, as they thought. It so happened that I was neither offended nor cross, just miserable."

Anne can render a moment in which everyone is talking simultaneously, acting or reacting, an example of barely contained chaos that poses a challenge for even the practiced writer. Conversely, there are tableaux that show us the characters in nearly static poses that communicate who they are, individually and collectively, and the levels of tension, resignation, or acceptance at which they have arrived.

Among such diary entries is one that John Berryman especially admired. Otto Frank is concerned about a business meeting taking place in the office downstairs. It's suggested that he listen in, with his ear to the floor. He does so, along with Margot, all morning and into the afternoon, until, half paralyzed—the man is in his midfifties—Otto gets up. Anne takes his place, but the drone of voices puts her to sleep and she wakes up having forgotten every word she's overheard.

Writes Berryman, "I have seldom, even in modern literature, read a more painful scene. It takes Anne Frank, a concise writer, thirteen sentences to describe."

Some of the most dramatic incidents are the real and false alarms, the actual break-ins as well as the frights occasioned by noises the workers and helpers make, little thinking the sounds might be mistaken for the arrival of the secret police. Dated April 11, 1944, the longest entry in the published version of the diary concerns a break-in. A peaceful domestic scene (a Monopoly game, a visit with Peter, an argument with Pfeffer over a cushion) is interrupted: someone is breaking in downstairs. The men surprise the thieves in the act of knocking a hole in the wall. Pretending to be the police, they scare the intruders away and cover the hole with a plank, but someone kicks it in from outside. A passing couple shines a flashlight into the opening, lighting up the warehouse. Silence, then more noise downstairs. Then silence again. The residents are left in the dark and cold; fear plays havoc with their stomachs, so that everyone has to use the lavatory. In the fetid atmosphere, the Jews wait until the helpers return and tell them how serious the danger has been. Police had come to investigate the burglary, but left without suspecting that Jews were upstairs.

Obviously, the diary entry was written after the crisis had ended. Yet by this point, Anne's narrative ability is so highly developed that she can re-create the terror into which she and the others were plunged as if she were still experiencing it, and without the mediating effect of knowing that the incident had a (relatively) happy ending.

There are comical interludes, such as the breaking of the sack of beans Peter is carrying upstairs. At first Peter is frightened, but then begins to laugh when he sees Anne standing at the bottom of the stairs, "like a little island in the middle of a sea of beans. I was entirely surrounded up to my ankles in beans." Anne and Peter attempt to gather the beans, which elude them, rolling into the corners and the holes in the floor. "Now, every time anyone goes downstairs they bend down once or twice, in order to be able to present Mrs. Van Daan with a handful of beans." In *Anne Frank Remembered*, Miep Gies recalls Otto Frank, returning to the office for the first time after the war, bending to pick up a bean.

Occasionally, horror is commingled with comedy, again in ways that deepen our understanding of Anne's "characters" and their interrelations. One night, Anne hears a sound so loud she fears that an incendiary bomb has fallen nearby. The Franks go upstairs to find the Van Pelses watching a red glow outside the window. Mrs. Van Pels is convinced that the warehouse has caught fire. The residents return to their beds—only to be awakened by more shooting:

"Mrs. Van Daan sat bolt upright at once and then went downstairs to Mr. Dussel's room, seeking there the rest which she could not find with her spouse. Dussel received her with the words, 'Come into my bed, my child!' which sent us off into uncontrollable laughter. The gunfire troubled us no longer, our fear was banished."

There are also memorable moments of reassuring domesticity. We watch, through Anne's eyes, the disorderly burlesque that results when Mr. Van Pels throws himself into a hands-on demonstration of his professional sausage-making expertise:

"The room was in a glorious mess. Mr. Van Daan was wearing one of his wife's aprons swathed round his substantial person (he looked fatter than he is!) and was busy with the meat. Hands smothered in blood, red face, and the soiled apron, made him look like a butcher. Mrs. Van Daan was trying to do everything at once, learning Dutch from a book, stirring the soup, watching the meat being done, sighing and complaining about her injured rib. That's what happens to elderly ladies (!) who do such idiotic exercises to reduce their large behinds."

Even more illuminating is the "potato-peeling scene," an episode Anne meant to stand alone as a short story, which she included in the office register in which she wrote the pieces published as *Tales from the Secret Annex*. Otto integrated some of these sketches into the diary, dating this one August 18, 1943. Anne focuses on those instants when character is revealed through the way a person deals with an object or objects—here, a peeler and a few potatoes.

As the chores are divided up—the setting out of the potatoes, the newspaper, and the pan of water—the temperature of the community is taken and its health diagnosed by Anne's observation that everyone keeps the best knife for himself. That Pfeffer is doing a terrible job doesn't prevent him from telling everyone else how to do it, nor from blustering in German when Anne ignores his advice.

Scowling, Otto concentrates as if his life, as if *everyone's* life, depends on his not producing even one "imperfectly scraped potato." Mrs. Van Pels tries to flirt with Pfeffer, then gets frustrated and bored, and starts picking on her husband. He's getting his suit dirty, making a mess. Doesn't he want to sit down? He makes compliant noises, but he's tuning her out. So she raises the ante, turning to a more fraught subject, the progress of the invasion. The English aren't flying as many bombing raids as they used to. Her husband blames the weather, but she replies that the weather's been fine. Then she adds that she's noticed that, unlike her husband, Mr. Frank always answers *his* wife when *she* speaks.

It's no longer about potatoes. The real subject is the Allied invasion. The Germans might still win. Also at issue is the Van Pelses' marriage: how long it can survive in the attic and whether they will survive at all. The British do nothing, says Mrs. Van Pels, and her husband yells at her to be quiet, slipping into German. *Enough.*

Mrs. Frank tries not to laugh. Anne looks straight ahead. They know it's only theater. When the Van Pelses are seriously fighting, they're quiet and careful with each other.

Anne's ability to dramatize becomes even more important when she begins to write about her romance with Peter. In January 1944, she records a conversation with Margot and Peter that continues when she is alone with him—a talk about the gender of Boche, the cat. Anne thinks that the cat is pregnant but is soon persuaded that it has gotten fat from a diet of stolen bones. In Anne's first draft, Peter insists that he's seen the cat having intercourse. There follows a fairly clinical description of animal castration, and a touchingly muddled attempt to determine the terminology for male and female genitalia.

The second, shortened and changed version has Peter telling Anne that, while playing with the cat, he noticed that "he's quite a tom." In both drafts, Anne prides herself on being able to have such a relaxed and grown-up talk with Peter about such a risqué subject; it proves what close, trusting friends they have become. But she may have had second thoughts about how future readers of *Het Achterhuis* might react. Peter's account of watching the cat have sex, and the mechanics of castration, are omitted from Anne's revision. She did, however, keep the observation that the impromptu biology lesson has left her feeling "a little funny," and that she finds herself replaying the scene in her mind:

"I wasn't quite my usual self for the rest of the day though, in spite of everything. When I thought over our talk, it still seemed rather odd. But at least I'm wiser about one thing, that there really are young people—and of the opposite sex too—who can discuss these things naturally without making fun of them."

The ease of the conversation about feline physiology is typical of Anne's approach to the subject of sex. Her openness has resulted in the book's being banned from schools and libraries, but in fact it's part of what young readers find refreshing, informative, and comforting. Her tracking of the highs and lows of her erotic preoccupation are still among the most accurate accounts of what it feels like to be a confused, romance-obsessed teen.

Late in the diary, Anne describes mining the Bible for its bewildering sexual information, reflecting on the scene in which the elders spy on Susanna in her bath, and wondering what exactly is meant by the guilt of Sodom and Gomorrah.

When I was growing up in the 1950s, the subject of sex was so much more veiled than it is today that I remember reading the Old Testament (Lot and his daughters! Boaz and Ruth!) for its tantalizing if vague allusions to sex. A friend confessed that when she read Anne Frank's diary as a preadolescent, she thought it was *all* about sex. Now that the air we breathe is so heavily saturated with eros that a child can learn the facts of life from an afternoon of talk shows and soap operas, it seems unlikely that the diary could teach kids something new about sex, except in so far as any kind of nonhysterical honesty about the topic is always new.

One striking aspect of the diary is how much life it packs into its pages. Sex is part of it, as is death, love, family, age, youth, hope, God, the spiritual and the domestic, the mystery of innocence and the mystery of evil.

In addition to all that, the diary is about Hitler's war against the Jews, about Holland during World War II, and about the Allied invasion of Europe as seen from inside an occupied country. It's easy to overlook the amount of history folded into these entries: "Saturday March 27, 1943. Rauter, one of the German big shots, has made a speech. 'All Jews must be out of the German-occupied countries before July 1. Between April 1 and May 1 the province of Utrecht must be cleaned out (as if the Jews were cockroaches). Between May 1 and June 1 the provinces of North and South Holland.' These wretched people are sent to filthy slaughterhouses like a herd of sick, neglected cattle."

As the hidden Jews followed the progress of the Allied invasion, Anne registers the statistics of each military maneuver that could be reported over the contraband radio. On D-day, she writes that 11,000 planes have been flying back and forth, bringing troops behind enemy lines, while 4,000 boats are ferrying soldiers and supplies between Cherbourg and Le Havre. The possibility that the Allies may be victorious in 1944, she writes, is a reason for fresh hope and, after everything they have endured, an inspiration to remain brave and calm. "Oh, Kitty, the best part of the invasion is that I have the feeling that friends are approaching. We have been oppressed by those terrible Germans for so long, they have had their knives so at our throats . . . now it doesn't concern the Jews any more; no, it concerns Holland and all occupied Europe."

Sadly, the war still very much concerned the hidden Jews. Yet during the last months in hiding, Anne and the others began to let themselves imagine they might prevail. They lost by the narrowest of margins—lost in Amsterdam, in Westerbork, in Auschwitz, and then again in Bergen-Belsen—as the luck that had kept them safe for two years turned against them.

Anne's diary is a symphonic composition of major and minor themes, of notes and chords struck at sufficiently regular and frequent intervals so that they never leave the reader's consciousness for very long. It's possible to trace each thread as it weaves through the diary, periodically reappearing to heighten and sharpen our understanding of a character or situation.

How amazing, a casual reader might say, how thoroughly unlikely that such a penetrating, dramatic, and structurally ambitious work should have evolved, on its own, from the natural and spontaneous jottings that a young girl added, every day or every few days, to her diary. Such a reader would have been right, or partly right, to wonder about that naturalness and that offhand improvisatory spirit.

FIVE

The Book, Part III

INCLUDED IN *THE CRITICAL EDITION* ARE NUMEROUS PHOTOS documenting the childish printing of Anne's first diary entries and the fluid cursive scrawled over the final pages. Forensic handwriting experts engaged by the Netherlands Institute for War Documentation have charted the alterations in each up-swing and loop as her handwriting developed from that of a child into that of an adolescent. But even more pronounced than the changes in penmanship are the differences in maturity and sensibility that separate the little girl who printed those awkward letters from the young author who covered the colored sheets with confident script.

The gap between the giddy River Quarter social life that Anne describes at the start of the diary and the self-searching meditations that conclude it is so immense that it has distracted readers and critics from something they might otherwise have noticed had they been thinking more clearly—that is, had they taken the diary more seriously. Though the *content* of the final pages is appropriately more somber and mature than that of the first entries, the *style* and the *voice* of the diary don't change all that much from the diary's beginning to its end.

The explanation is that, as we have seen, Anne rewrote the early sections two years after their initial composition, and that Otto combined the first draft with her revisions to produce a manuscript that told Anne's story in the most affecting and *consistent* way. Had we considered the differences in the ability to comprehend and articulate that one would expect from the most precocious thirteen-year-old and the intellectual and literary capabilities of a girl of fifteen, we might have concluded (even in advance of the 1986 *Critical Edition*) that the diary was not precisely what it seemed.

Indeed, a glance through *The Critical Edition* confirms how much work Anne herself—and later her father—did on the manuscript. Had the original diary not been edited by Anne and Otto, it seems less likely that it would have been published. Readers might have been put off by the sections that were (unlike many of the passages that open the diary, but which were written after the dates under which they appear) *actually* written when Anne had just turned thirteen. Here, for example, is the list of birthday gifts from the checkered diary, which does not appear in *The Diary of a Young Girl*:

"From Mummy and Daddy I got a blue blouse, Variety, which is the latest party game for adults, something like Monopoly, a bottle of grape juice, which to my mind tasted a bit like wine and which has now begun to ferment so that I can't drink it any more and I may have been right, since wine is made from grapes after all; then a puzzle; a bottle of peek-aroma 'with acorns' (I got that later, I mean 'the acorns'); a jar of ointment; a 2 ½ guilder bank note; a token for two books; a book from Katze, the Camera Obscura, but Margot has got that already, so I swapped it."

Though we can do without most of this, it might have been instructive to learn that Anne received gifts from Peter van Pels and from Fritz Pfeffer's "wife." But it is only interesting *after* we've read the book, and it's hard to imagine even the diary's most devoted fans working their way through this, or any, of the bubbly longueurs of the "a" version.

Indeed, the first pages of the red, gray, and tan checked book are full of jottings that would have dissuaded most adults—and most children, for that matter—from continuing. In passages that both Anne and her father cut from *Het Achterhuis*, Anne lists her teachers for every grade and her daily activities. An inventory of her classmates annotated with gossipy commentary engages us, if at all, only because it reveals a bit more about Anne's personality during that relatively carefree time. "Miss J. always has to be right. She is very rich and has a wardrobe full of gorgeous dresses, but they're much too old for her. . . . Henny Mets is a nice, cheerful girl, except that she talks much too loudly, and is very babyish when she plays in the street. . . . Rob Cohen was also in love with me, but now I can't stand him any more he is a hypocritical, lying, whining, crazy, boring little boy, who thinks he's the cat's whiskers."

Months after life in Amsterdam had become nearly untenable for the city's Jews, and five days before her sister received the decisive call-up notice, Anne wrote an unadorned account, which she cut and which Otto did not restore, of a trip to an ice-cream parlor, Oasis, with two boys, Hello and Fredie, and a girl named Wilma:

"We went to oasis and bought an ice cream for 12 cents, then Wilma came, and they wanted to stand us another ice cream but Wilma and I didn't want to, but they bought us each one for 12 (cents) anyway, but we didn't accept it, so Fredie and Hello had 2 more 12-cent ice creams."

Anne Frank was a prodigy, but her gifts had not yet developed, at thirteen. The evidence of those gifts would come only later, brought about at least partly by what John Berryman called "the special pressure" of her incarceration in the secret annex.

ANNE intended *Het Achterhuis* to begin with the June 20, 1942, entry, in which she decides to write to Kitty and wonders who will be interested in the "unbosomings of a thirteen-year-old schoolgirl." The passage was composed at some point during the spring of 1944. Anne was about to turn fifteen and was revising her journal, recalling—and writing as—her thirteen-year-old former self, a popular girl with lots of boyfriends surreptitiously peeping at her in the mirrors on the classroom walls. She describes sitting with her chin in her hands, too enervated and bored to decide whether or not to stay home or go out. In reality, the decision about whether to go outside had been made for her two years before, and thinking her way back to that point required such an effort of the imagination that, at least once, Anne overdoes it and begins to gush, referring in the revision to "the dearest darling of a father I have ever seen." Sagely, Otto simplified it to "father" in the published version.

In another entry, dated Sunday, June 21, 1942, and also written two years afterward, Anne reflects on the surprising notion that a girl of her age should be so obsessed with boys. For the sake of *Het Achterhuis*, she pretends to still be the flirty chatterbox puzzling over how to discourage overeager suitors. But here too she is remembering what it was like to allow a boy to walk her home from school and to expect him to fall madly in love with her. When a boy blows kisses at her or tries to take her arm, Anne tells Kitty, she gets off her bicycle, pretends to be insulted, and orders him to leave her alone. Even if Anne *had* been writing this on June 21, 1942, the entry would still have been outdated. Nine days before, all Jews had been required to turn in their bicycles.

In this ghostly collaboration between the living and the dead, Anne and her father seem to have agreed that her diary should start with a sketch of how she lived before she vanished into the attic. But the breathless tone in which the diary, as we now know it, opens—"On Friday, June 12th, I woke up at six o'clock, and no wonder; it was my birthday"—was not how Anne wished her readers to first hear her voice. Revising, Anne excised the description of her joy at receiving the diary and of the birthday party that featured a Rin Tin Tin film.

Otto chose wisely in restoring these accounts of childish pleasures enjoyed in freedom. But with the exception of those few paragraphs rescued from the original draft, the diary's early pages are the work of the fifteen-year-old Anne writing more thoughtfully than she could have two years before. Anne herself retained—and expanded upon—the references to her Ping-Pong club, her trips to the ice-cream parlor, to the anxiety that grips her fellow students before a teacher's meeting, and of being assigned to write an essay as punishment for her excessive chattering.

If the innocent thirteen-year-old sets off by listing her presents, and noting which friends gave her which gifts, the fifteen-year-old begins by reflecting on the oddness of a girl like herself keeping a diary. As Anne knew, plenty of girls keep diaries. More to the point, even as Anne was asking Kitty who would want to read about her life, she understood that, given the extraordinary circumstances under which she was living, the idea of *her* keeping a diary wasn't odd at all. Even as she is writing that she doesn't "intend to show this cardboard-covered notebook" to anyone, she is not only revising a book that she hopes to get published, but she is no longer writing in the cardboard-covered diary except to fill in blank pages.

Aware that strangers might read *Het Achterhuis*, realizing that she needed to explain how "Anne Robin" came to be confined to the house behind, she added the section in which her father suggests they might have to go into hiding, reproducing the essence if not the letter of their conversation. One can understand why Anne chose not to record this scene soon after it happened; perhaps this grim possibility seemed too alarming to dwell on. Only later, for clarity, does she return to the moment when she heard about the plan that might shortly be put into effect. In the revised version, Otto tells Anne that they want to avoid being arrested and having their possessions seized by the Germans. So they have decided to disappear before they are tracked down and deported.

"'But, Daddy, when would it be?'

"He spoke so seriously that I grew very anxious.

"'Don't you worry about it, we shall arrange everything. Make the most of your carefree young life while you can.'

"That was all. Oh, may the fulfillment of these somber words remain far distant yet," is how Anne ends the scene, though the truth is that, by then, the warning of those somber words had already been fulfilled.

Hoping that she is telling a story that will interest, among others, the Dutch minister of education in exile, she writes the June 20 entry as an introduction to herself and to Kitty. Promising "to bring out all kinds of things that lie buried deep in my heart," she describes a feeling that will be familiar to many, perhaps most, adolescents—the sense of being solitary even when she is surrounded by friends and loved ones. By the time she is describing her alienation, she *is* alone, or at least cut off from the "thirty people whom one might call friends." Her diary has become a weapon in her struggle against the isolation that has increased in the absence of companions her own age, and despite (or exacerbated by) the close quarters in which she and her family and the others have been living.

Here is Laureen Nussbaum's commentary on the June 20 diary letter: "Anne, putting herself in her state of mind of two weeks before she went into hiding, explains why, despite all her popularity, she feels lonely and in need of a true friend to whom she can direct her outpourings. That friend she decides to call Kitty and after a terse version of her original autobiographical sketch, she proceeds immediately to write her first 'Dear Kitty' epistle. In just four letters she summarizes both her school and her social life in the spring of 1942 and ends with a beautiful transition: an evening stroll with her father, during which he broaches to her the subject of hiding and all the drastic change which the move will entail."

Putting herself in her state of mind of two weeks before she went into hiding. That is precisely what Anne is doing—that is how memoirs are written, in this case in the form of a diary, or a series of letters. Anne was not trying to fictionalize but rather to give the most accurate chronological record of the person she was and the person she became, and of everything and everyone that helped bring about that change.

IN GENERAL, comparisons of the first entries and their counterparts in the second draft persuade us that Anne was right to trust her instinct for self-editing. Typically, the revised version is clearer, more readable, and free of the sketchiness and haste that muddle some early passages. The differences between Anne's initial efforts and her revisions vary from trivial to profound, and deepen our respect for her as a writer. The first versions are in many cases more impulsive and spirited, the second more distanced, cooler, even abstract. The revisions may trade immediacy for clarity, raw emotion for reflection, but they are nearly always better *written*—more condensed, descriptive, fully dramatized, and evocative. Only very rarely, when Anne overthinks her own reactions to events, does the writing become more literary or "interesting" in ways that seem less faithful to what she might have thought and felt at the moment.

In Anne's original March 12, 1944, entry, we learn that she has just heard about the arrest of someone she knows and about the illness of Bep's father—bad news that makes her want to fall asleep as a release from thinking. After a reference to her isolation and to the divide between her inner and outer selves, she explains why she is unable to turn for help to her sister: "Margot would so much like to be my confidante, but I can't. She's a darling, she's good, she's pretty, but she lacks something I need. Nor could I bear to have someone about all day long who knew what was going on inside me. I can't have my confidant around me all day long, except for . . . Peter!"

The differences in the second version are subtle but all important. "Margot is very sweet and would like me to trust her, but still, I can't tell her everything. She's a darling, she's good and pretty, but she lacks the nonchalance for conducting deep discussions; she takes me so seriously, much too seriously, and then thinks about her queer little sister for a long time afterwards, looks searchingly at me, at every word I say, and keeps on thinking: 'Is this just a joke or does she really mean it?'"

Only in this draft has Anne found the phrase ("she lacks the nonchalance for conducting deep discussions") that Philip Roth singled out as indicative of her complexity of mind and grace of expression. Charitably, Anne concludes, "I think that's because we are together the whole day long, and that if I trusted someone completely, then I shouldn't want them hanging around me all the time." The mention of Peter as the ideal confidant—the outburst that ends the first draft—has disappeared completely.

WHAT we have of Anne's revisions—that is, the "b" version—
ends in March 1944. The entries dated after that exist only in
one draft, so one cannot know how much she intended to rewrite or retain of the sections that describe the intensification
of her involvement with Peter. But every reference to Peter that
she *did* revise (presumably, after her infatuation had cooled)
is toned down so that the romantic becomes platonic, and the
strong emotions seem, by contrast, neutral, the sorts of things a
girl might say about any close neighborhood friend.

Early in 1944, Anne describes the thrill of gazing into Peter's eyes. "I couldn't refrain from meeting those dark eyes
again and again, and with my whole heart I almost beseeched
him, oh, tell me, what is going on inside you, oh, can't you
look beyond this ridiculous chatter?" The revision is notably
less heated: "It gave me a queer feeling when I looked into his
deep blue eyes and saw how embarrassed this unexpected visit
had made him.

"I would have liked to ask him: Won't you tell me something about yourself, won't you look beyond this ridiculous
chatter? . . . When I lay in bed and thought over the whole situation I found it far from encouraging and the idea that I should
beg for Peter's patronage was simply repellent."

The February and March 1944 entries in the "a" version are
cut to eliminate impassioned references to Peter and speculations about whether Anne's feelings are returned. In the "a"
version, Anne tells Kitty that "From early in the morning till
late at night, I really do hardly anything else but think of Peter.
I sleep with his image before my eyes, dream about him and he
is still looking at me when I awake." Missing from the revisions,
this passage reappears in Otto's edit of *The Diary of a Young Girl*.
In the first draft, and the published diary, Anne writes that she
lives from one meeting with Peter to the next, but that obses-

sion no longer haunts the writer editing *Het Achterhuis* for publication.

On March 7, there is a revealing alteration. The original reads: "At the beginning of the New Year the second great change, my dream . . . and with it I discovered Peter . . . discovered my longing for a boy; not for a girl friend but for a boy friend. I also discovered my inward happiness and my defensive armor of superficiality and gaiety. Now I live only for Peter, for on him will depend very much what will happen to me from now on!"

Anne's rewritten dream of her future is nobler, more abstract—and no longer dependent on Peter:

"At the beginning of the New Year the second great change, my dream . . . and with it I discovered my boundless desire for all that is beautiful and good." Nowhere in the revisions do we find passages like the following, which exists only in the first draft: "Peter has touched my emotions more deeply than anyone has ever done before—except in my dreams. Peter has taken possession of me and turned me inside out." Nor did Anne choose to include the discussions about sex that the "a" version records—conversations in which Peter explained how contraceptives function and Anne informed him about the mysteries of female anatomy.

Even small fixes are telling. In the March 12, 1944, entry, Anne enumerates the crises that make her long to fall asleep and wish she could confide in her sister; her worries include the fact that she never receives a "friendly glance from Peter." In the second draft, Anne admits that she loves talking to Peter, though she fears being a nuisance. "But still, I won't drive myself mad over it, I see quite a lot of him and there is no need to bore you with it too, Kitty, because I'm miserable. On Saturday afternoon I felt in such a whirl after hearing a whole lot of sad news that I went and lay on my divan for a sleep. I only wanted to sleep to stop thinking." Again, there is no hint that Peter's coolness is one of the subjects about which Anne longs to stop thinking.

Laureen Nussbaum observes, "When revising her black notebook during the late spring and early summer of 1944, only a few months after she had filled it with her outpourings, Anne had become very critical of her infatuation with Peter van Pels . . . in the *b* manuscript she eliminates most of her effusive entries of that emotional period . . . Otto Frank reinstated the bulk of those eliminations." Otto "selected time and again the more emotional passages of Anne's *a* version, some of which Anne had dispensed with, while she had reworked others into fictional stories. By the time she was rewriting her entries of the beginning of 1944, Anne had gone through a great deal of inner development. Father Frank ignored all of that evidence of growth. Or did he want to preserve a stormy stage in the development of his beloved little Anne rather than allow her to present herself as the more objective and self-contained young writer she had become at such a precocious age? One can only speculate."

What is there to speculate about? Otto's restitution of these cuts created a more compelling drama. Had Otto expunged Anne's romance with Peter, Broadway and Hollywood would likely have wanted to reinstate or invent it. And it wasn't as if Otto *created* the sections in which his daughter tells Kitty about her nascent love affair in the way that an adolescent girl would confide, in her best friend, the details of her first serious crush. But after the onset of her disappointment in Peter, Anne did not imagine the heroine of *Het Achterhuis* as a lovesick teen, agonizing over every smile she got from the boy upstairs.

Two passages from the same entry that describes Minister Bolkestein's radio speech typify the differences between successive drafts. In both, Anne's focus shifts from inside the attic, where the scared "ladies" wait out the air raids, to the wider world outside, so that future generations can see, as the minister suggested, how the Dutch people suffered.

In the first version, Anne reports what she has heard and read about the deterioration of civil society. More than four years into the Nazi occupation, the Dutch are now enduring the additional hardship of Allied bombings:

> . . . how the houses shake from the bombs, how many epidemics there are, such as diphtheria, scarlet fever etc. What the people eat, how they line up for vegetables, and all kinds of other things, it is almost indescribable.
>
> The doctors here are under incredible pressure, if they turn their backs on their cars for a moment they are stolen from the street, in the Hospitals there is no room for the many infectious cases, medicines are prescribed over the telephone.
>
> Above all the countless burglaries and thefts are beyond belief. You may wonder whether the Dutch have suddenly turned into a nation of thieves. Little children of 8 and 11 years break the windows of people's homes and steal whatever they can lay their hands on, you can't leave your home unoccupied, for in the five minutes you are away your things are gone too.

Here is the same account, from the draft Anne revised:

> . . . how the houses trembled like a wisp of grass in the wind, and who knows how many epidemics now rage . . . People have to line up for vegetables and all kinds of other things; doctors are unable to visit the sick, because if they turn their back on their cars for a moment they are stolen; burglaries and thefts abound, so much so that you wonder

*what has taken over the Dutch for them suddenly to have
become such thieves. Little children of eight and eleven years
break the windows of people's homes and steal whatever they
can lay their hands on. No one dares to leave his house unoccupied for five minutes, because if you go, your things go too.*

Admittedly, there's an added bit of *writing*. Explosions may
have shaken Holland, but did the houses really tremble like
wisps in the wind? But every other large and small change is for
the better. To say that epidemics rage is stronger than simply
enumerating them. Words and phrases that writers are sensibly
advised to avoid—"indescribable," "incredible," and "beyond
belief"—have been eliminated or replaced by more descriptive
adjectives.

The detail of lining up for food has been selected from more
vague ones, and the doctors' problems have been distilled to the
inability to make house calls without their cars being stolen.
Minor alterations increase a sentence's effect on the reader.
Compare "You can't leave your home unoccupied for in the five
minutes you are away your things are gone too" with "if you go,
your things go too."

Considering that Anne began her revisions in the spring of
1944 and that by August the family had been arrested, the above
passage had to have been rewritten within a short time. Only a
few weeks, months, or hours separated Anne's two drafts.

The differences between the drafts are naturally more pronounced when more time has elapsed between them. Some
revisions are sobering corrections based on a less optimistic
awareness of how things would turn out. On September 21,
1942, when Anne contemplates her first winter in hiding, she
writes that all her warm sweaters have been left with friends,
but that Miep may ask if she can store them for Anne, who will
then get the sweaters back. By the time she is rewriting, this

seems not to have happened: "We have some clothes deposited with friends, but unfortunately we shall not see them until after the war, that is if they are still there then."

Passages of remembered dialogue are altered and clarified. Near the diary's beginning, several members of the household are speaking, observed by the others. Anne Frank finesses the scene, and then (in the second draft) improves it further.

It is late September, almost three months after the two families have gone into hiding, long enough for them to have begun getting on each other's nerves. Hermann van Pels tells Anne that it's self-sabotaging to be overly modest. In part, he's reacting to Otto Frank's having just been praised for his modesty, but he's also giving advice the way adults often do to children: reflexively, without bothering to notice who the child is, or to consider what advice she could actually use. The briefest acquaintance with Anne should have alerted Mr. Van Pels to the fact that modesty was not her problem.

The first version of the scene does a perfectly adequate job of re-creating a conversation in which more is unspoken than expressed, and in which at least two of the speakers enjoy a foray into passive aggression:

> One Sunday morning we were sitting at breakfast, and we were talking about how modest Daddy is and then Mrs. v.P. said:
>
> "I too have an unassuming nature, more so than my husband!"
>
> Mr. v.P.: "I don't wish to be modest," and to me: "Take my advice, Anne, don't be too unassuming, it will never get you anywhere!" with which Mummy agreed.
>
> Mrs. v.P.: "What a stupid thing to say to Anne, that outlook on life is just too silly!"
>
> Mummy: "I myself also think that you don't get much

*further with it. Just look, my husband and Margot and Peter
are exceptionally modest while Anne, your husband, and I
are not modest at all. We are not immodest, but we are not
modest either."*

*Mrs. v.P.: "Oh, no, on the contrary, I am very modest,
how can you say that I am immodest?"*

*Mummy: "I haven't said that you are immodest, but you
aren't all that modest either."*

But it's in the second attempt that Anne succeeds in giving
the moment its full complexity, animation, and humor:

*Somehow or other, we got on to the subject of Pim's ex-
treme modesty. Even the most stupid people have to admit
this about Daddy. Suddenly Mrs. v.P. says, "I, too, have an
unassuming nature, more so than my husband."*

*Did you ever! This sentence itself shows quite clearly how
thoroughly forward and pushing she is! Mr. v.P. thought he
ought to give an explanation regarding the reference to himself. "I don't wish to be modest—in my experience it does not
pay." Then to me: "Take my advice, Anne, don't be too unassuming, it doesn't get you anywhere"; Mummy agreed with
this too. But Mrs. van Pels had to add, as always, her ideas
on the subject. Her next remark was addressed to Mummy
and Daddy. "You have a strange outlook on life. Fancy saying
such a thing to Anne; it was very different when I was young.
And I feel sure that it must still be so, except in your modern
home." This was a direct hit at the way Mummy brings up
her daughters.*

*Mrs. v.P. was scarlet by this time. Mummy calm and cool
as a cucumber. People who blush get so hot and excited, it is
quite a handicap in such a situation. Mummy, still entirely
unruffled, but anxious to close the conversation as soon as*

possible, thought for a second and then said: "I find, too, Mrs. v.P., that one gets on better in life if one is not overmodest. My husband, now, and Margot, and Peter are exceptionally modest, whereas your husband, Anne, you, and I, though not exactly the opposite, don't allow ourselves to be completely pushed to one side." Mrs. v.P.: "But, Mrs. Frank, I don't understand you; I'm so very modest and retiring, how can you think of calling me anything else?" Mummy: "I did not say you were exactly forward, but no one could say you had a retiring disposition."

THE REVISION not only displays a greater facility, but a sharper sense of who her relatives and roommates are and of how they see themselves and each other. Anne's ear is more attuned to the way they speak, and, more important, to what they mean. Two years after the event, she hears their voices more clearly, even as the redrawn portrait of Mrs. Van Pels shows the wear and tear of her neighbor's perpetual insistence on having the last word.

In the second version, Anne provides cues for her actors— the coolness, the blushing—and makes sure we notice Mrs. Van Pels's vulgarity. Nor, this time, can we miss the fact that Mrs. Van Pels's apparently offhand remark about modernity is a dig at Mrs. Frank's child-rearing practices, a veiled insult that doesn't quite show through, when, in the earlier draft, she merely comments on the silliness of "that outlook on life." Mrs. Frank's reply is not only more elegant, but funnier, slyer, and more pointed. Cleverly, she groups herself with the "immodest," so that no one could accuse her of singling out Hermann and Auguste van Pels for special criticism. Unlike Hermann, Edith knows there is no point in even pretending that Anne is modest; indeed, Anne's confident self-regard seems to have been one of the differences that generated friction with her mother.

Although we are meant to be on Mrs. Frank's side of this altercation, the calm double-edgedness of her responses alert us to what a difficult opponent she could be, as indeed Anne found her. The tweaks that turn Mrs. Frank's rejoinder into the more polite and frostier response could serve as an example of why a writer revises, and of the difference a few words can make.

Mrs. Van Pels goes on to defend herself: if she weren't pushy, she would starve to death. It bears repeating that the two families have been living together for only three months, and are simultaneously realizing the importance and the difficulty of remaining civil. Already there is trouble about food—shortages and rationing—about how much each resident consumes, tensions that can arise even in close families, when there is plenty to eat. It's a freighted moment, and Mrs. Van Pels is being intentionally provocative when she suggests that they are involved in a Darwinian struggle requiring aggression and perseverance.

Mrs. Frank laughs—more, we assume, from discomfort than because Mrs. Van Pels has said something funny. The furious Mrs. Van Pels sees Anne shaking her head and gets even angrier. In the earlier draft, she "delivers another sermon," but in the second she lets loose with "a lot of harsh German, common, and ill-mannered, just like a coarse, red-faced fishwife—it was a marvelous sight."

By the end of the scene, Mrs. Van Pels's defensiveness and peevishness have burbled to the surface, leaving a kind of residue that will slick whatever we read about her from that point on. It has been argued that casting Shelley Winters as Petronella van Daan in the film was such an inspired choice that Winters's high-strung, flirtatious crudeness, as well as the fragility and terror underneath, forever formed our image of Otto's business partner's wife. But Winters's performance, however inspired, is only a veneer, layered over what already exists in the diary.

Thousands of people died during the forced marches that

followed the evacuation of Auschwitz, but Auguste van Pels's name is one of the few, or only, ones we know. And all because of a diary in which a young girl recorded a petty argument in which the older woman could hardly have seemed more irritating—or more human. Among the reasons we remember her is a single instant, unrepeatable in time, when she sees a little girl shaking her head, and explodes. Anne may have judged her neighbor and exposed her frailties and flaws. But she also made her live on the page, thus allowing the facts of history and the passage of time to act upon, and soften, the severity of that judgment.

BY AND large, the most useful revisions increase the clarity of the text. Confusing descriptions are sorted out and reordered, necessary facts added. In September 1942, there is a "big drama" in the "a" version; by the time Anne is editing, she has lived through enough *real* dramas to make the event seem more like "a little interruption in our monotonous life." Margot and Peter, who are allowed to read everything, have been forbidden a certain book. Originally, it's a book "about the last war" that Mr. Kleiman has brought to the attic. In the revision, Anne specifies that it is a book "on the subject of women."

    This explains the blowup that ensues, as well as the comedy of Mr. Van Pels grabbing the proscribed volume from his son and protecting the young peoples' innocence by keeping it himself. As Sylvia P. Iskander explains in an essay about Anne's reading, the controversial work was Jo van Ammers-Küller's *Heren, knechten, en Vrouwen* (*Gentlemen, Servants, and Women*). "In this first book of a trilogy about the *burgemeester*, or mayor of Amsterdam, the mayor considers betraying his country's alliance with England by assisting the French in sending arms to the American colonies in their fight for independence. Whether the issues of patriotism, betrayal, or sex, or all of them made the

Franks temporarily censor the book for their thirteen-year-old is impossible to say." In any case, Anne was allowed to read the book a month later, and after another year or so, Anne's parents let her read nearly everything she wanted. ·

In October 1942, Anne's original entry makes it tricky to determine the chronology of events that led to a "terrible fright." It requires effort to figure out that the incident begins when a noise is heard on the steps. First Anne thinks it's Mummy or Bep, and then it turns out to be the carpenter, whose presence on the stairs traps Bep in the attic. Someone rattles the door, there's whistling and thumping. Anne and the others are sure that the carpenter has found them out, but it turns out to be Mr. Kleiman, and they can relax.

On the second attempt, Anne not only gets it right, but lets the reader of *Het Achterhuis* know that the carpenters have come to fill the fire extinguishers. "Downstairs they are such geniuses" that no one has informed them about the workmen's scheduled visit, and when the attic residents hear men on the stairs, a silence falls over their rollicking lunch with Bep. The anxious moment is dramatized so that now we see it from the perspective of the hidden Jews and their helper, and we are still in their viewpoint when the confusion is cleared up.

"After he'd been working for a quarter of an hour, he laid his hammer and tools down on top of our cupboard (as we thought!) and knocked at our door. We turned absolutely white. Perhaps he had heard something after all and wanted to investigate our secret den. . . . The knocking, pulling, pushing, and wrenching went on. I nearly fainted at the thought that this utter stranger might discover our beautiful secret hiding place. And just as I thought my last hour was at hand, we heard Mr. Kleiman's voice say 'open the door, it's only me.'" The revision goes on to explain that the hook that holds the swinging bookcase concealing the annex door had gotten jammed. So now both the chronology and the causes, hard to follow before, are unmistakable.

In the process, Anne changed her account of what went through her mind during those long minutes of uncertainty. In the earlier version, "I saw us all in a concentration camp or up against a wall." In the revised draft, her thoughts have turned from herself to the intruder, from the fates threatening her and the others to a more literary personification of evil. It's one of the rare instances in which the original diary is more persuasive than is the would-be author of the *Joop der Heul*–style thriller *Het Achterhuis*: "In my imagination the man who I thought was trying to get in had been growing and growing in size until in the end he appeared to be a giant and the greatest fascist that ever walked the earth."

Writing purely for herself gave Anne the freedom to assume that every reference would be understood, but writing for others necessitated explanation. In the first version, Mr. Van Pels alludes to the trick that the Franks used to make their neighbors think they had escaped Holland:

"Mr. van Pels repeated the story about Daddy being friends with an army captain who had helped him get away to Belgium, the story is now on everyone's lips and we are greatly amused."

In the second draft, Anne has Mr. Van Pels tell the story—which involves the Franks' landlord—in dialogue and in detail: "'I discovered a writing pad on Mrs. Frank's desk with an address in Maastricht written on it. Although I knew that this was done on purpose, I pretended to be very surprised and shocked and urged Mr. G. to tear up this unfortunate little piece of paper without delay. I went on pretending that I knew nothing of your disappearance all the time, but after seeing the paper I got a brain wave. "Mr. G."—I said—"it suddenly dawns on me what this address may refer to. It all comes back to me very clearly, a high-ranking officer was in the office about six months ago, he appeared to be very friendly with Mr. F. and offered to help him, should the need arise. He was indeed stationed in Maastricht. I think he must have kept his word and somehow or other managed to take Mr. F. along with him to Belgium and then on to Switzerland. I should tell this to any friends who may inquire, don't of course mention Maastricht."'"

In her revisions, Anne added blocks of information to help the reader envision the daily rituals and the quarters in which the attic residents barely managed to keep out of one another's way. The floor plan of the annex doesn't appear in the original diary. Anne would hardly have needed to map, for herself, an architecture with which she was so familiar. But the diagram is useful for the reader wondering how a bachelor and two families divided their tiny space during the day, and reapportioned it at night.

The elaborate system that the residents work out for bathing appears in the second draft; the first version focuses on how *Anne* manages this challenge. The charming and informative "Prospectus and Guide to the Secret Annex"—an ironic list of attractive features ("beautiful, quiet, free from woodland surroundings, in the heart of Amsterdam") and house rules ("Residents may rest during the day, conditions permitting, as the directors indicate") that portrays the cramped attic as a luxury health spa, does not appear in the original diary, but is incorporated in the revised version; it is identified as a "v. P product," though Anne fails to explain which of the Van Pelses was responsible.

Another important distinction between the first version and the revision has to do with the development of Anne's spirituality. In her book *Anne Frank: A Hidden Life*, Mirjam Pressler tracks Anne's references to God, which begin to appear only after her terrors—occasioned by the break-ins, the bombings, her growing sense of isolation and doom—move her to seek solace in religion. Until November 1943, most of the references to God appear in the revisions but not in the original. But "on November 27, 1943, Anne writes about praying for the first time, and directly asks God to help her. From now on God and nature—seen as interchangeable—take on the function of comforting her, cheering her, and soothing her fears."

PERHAPS the most startling clarification of a single event occurs in the entry for Sunday, July 5, the day on which the Franks received the call-up order for Margot to report for deportation.

Anne's first account of the afternoon reflects the shock she was still feeling two days after her family's arrival in the annex. The early version makes it sound as if Anne hears the policeman asking for her sister, and it's hard to tell when and how Anne learned whom the call-up was really for.

> At about 3 o'clock a policeman arrived and called from the door downstairs, Miss Margot Frank, Mummy went downstairs and the policeman gave her a card which said that Margot Frank has to report to the S.S. Mummy was terribly upset and went straight to Mr. van Pels he came straight back to us and I was told that Daddy had been called up. The door was locked and no one was allowed to come into our house any more. Daddy and Mummy had long ago taken measures, and Mummy assured me that Margot would not have to go and that all of us would be leaving the next day. Of course I started to cry terribly and there was an awful to-do at our house.

In the second draft, it's much easier to track the events of the interval during which the Frank women waited for Otto to return from the Jewish hospital. It had taken Anne two years to be able to write lucidly about that day, yet she maintains, in the revision, the sense of emergency that consumed her family.

> *At three o'clock . . . someone rang the front doorbell. I was lying lazily reading a book on the verandah in the sunshine so I didn't hear it. A bit later, Margot appeared at the kitchen door looking very excited. "The S.S. has sent a call-up notice for Daddy," she whispered. "Mummy has gone to see Mr. van Pels already. . . . Of course he won't go . . . Mummy has gone to the v.P.s to ask whether we should move into our hiding place tomorrow . . ."* We couldn't talk any more, thinking about Daddy, who, little knowing what was going on, was visiting in the Joodse Invalide. [After Mummy and Mr. Van Pels return] *Margot and I were sent out of the room, v.P. wanted to talk to Mummy alone. . . . When we were alone together in our bedroom, Margot told me that the call-up did not concern Daddy but her. I was more frightened than ever and began to cry. . . . Margot and I began to pack some of our most vital belongings into a school satchel, the first thing I put in was this diary, then hair curlers, handkerchiefs, schoolbooks, a comb, old letters. I put in the craziest things with the idea that we were going into hiding, but I'm not sorry, memories mean more to me than dresses . . . At five o'clock Daddy finally arrived and we phoned Mr. Kleiman to ask if he could come round in the evening.*

Most of what we know about the Franks' walk from their apartment to the hiding place at 263 Prinsengracht comes from the additions that Anne made to her sketchy original entry. Soon after her arrival in the attic, she wrote this first account of how they had gotten there:

"We left the house by a quarter to eight I had a (combinashion) on then two vests and two pairs of pants then a dress and a skirt then a wool cardigan and a coat, it was pouring and so I put on a headscarf, and Mummy and I each carried a satchel under our arm. Margot went too with a satchel on her bicycle, and we all made for the office. Daddy and Mummy now told me lots of things. We would be going to Daddy's office and over it a floor had been made ready for us."

Two years later, that morning becomes a scene in a book—the event from which so much else will follow. Anne adds details, slows the narration, and contextualizes it so that there is no mistaking how everything happened, and why. In addition, there is a sprightliness to the prose, a vigor missing from the numbed, bare-bones outline of two years before. When Anne was revising, the Allied advance was under way, and Anne was probably hoping that the terrifying July morning might come to seem like the start of an adventure in a *Joop ter Heul* detective novel. *How amusing it would be for future generations.*

Luckily it was not so hot as Sunday. Warm rain fell all day. We put on heaps of clothes as if we were going to the North Pole, the sole reason being to take clothes with us. No Jew in our situation would have dreamed of going out with a suitcase full of clothing. I had on two vests, three pair of pants, a dress, on top of that a skirt, jacket, summer coat, two pairs, lace-up shoes, woolly cap, scarf, and still much more; I was nearly stifled before we started, but no one inquired about that. Margot filled her satchel with schoolbooks, fetched her bicycle and rode off behind Miep into the unknown, as far as I was concerned. You see I still didn't know where our secret hiding place was to be . . . Only when we were on the road did Mummy and Daddy begin to tell me bits and pieces about the plan. For months as many of our goods and chattels and our necessities of life as possible had been sent away, and they were sufficiently ready for us to have gone into hiding of our

own accord on July 16. The plan had had to be speeded up 10
days because of this call-up, so our quarters would not be so
well organized but we had to make the best of it. The hiding
place would be in the building where Daddy had his office.

As a boy of five, Mozart was already composing. Keats was
dead at twenty-six. Maturity and creativity are unpredictable
over a lifetime, and the early appearance of genius frequently
obliges us to rethink our preconceived notions of age. When I
read in a book review or hear in a writing class that a child of a
certain age would never have such a grown-up response or use
such a sophisticated expression, I find myself resisting.

Even so, I can't help thinking that the description of Miep
and Margot riding "into the unknown, as far as I was concerned" does not sound like a thirteen-year-old who, days
before, saw her sister threatened with deportation, and by the
next morning had left her house and her life and moved into an
attic. It was, it could only have been, the phrase of an older girl,
looking back.

IN 1995, the so-called *Definitive Edition* of the diary was published
to heated media attention, warmer than the relative chill that
had greeted *The Critical Edition* when it appeared in English six
years before. Except for the "missing five pages," all three drafts
of the entire diary were included in *The Critical Edition*. But the
publicity surrounding the more reader-friendly *Definitive Edition*
implied that this was the English-language reader's first chance
to learn more about Anne than her father had chosen to reveal.
The suggestion was that the hidden had finally been made public, and a certain amount of prurient interest was generated by
Anne's disquisition on female genitalia. "The little hole underneath is so terribly small that I simply can't imagine how a man
could get in there, much alone how a whole baby can get out."

In fact, the *Definitive Edition* begins with a foreword explaining that everything it contained was available in *The Critical Edition*. But even if you have *The Critical Edition* in front of you, it's confusing to follow the three overlapping narratives in parallel bands, and it's unsurprising that only a small number of scholars and critics (and probably fewer general readers) went to the trouble.

In an incisive essay, Laureen Nussbaum guides readers through the versions and revisions. She explains what Otto Frank did and didn't do, and the controversies that have erupted over each successive revelation. "Otto Frank had picked and chosen from Anne's extant diary versions when assembling the typescript on which the original (1947) edition and the subsequent translations into dozens of languages would be based. He had added some of the vignettes she had written separately about life in the back quarters, made several rearrangements and corrections, while omitting some passages which he deemed either too irrelevant or too personal to include. In other words, Otto Frank had edited his daughter's diary, to which he had, of course, a perfect right: a prefatory note to this effect, however, would have saved him many future problems."

The inclusion of Anne's reflection on female anatomy, as well as the fact that the pacing in the longer *Definitive Edition* is slower than in the earlier edition of *The Diary of a Young Girl*, are likely among the reasons that the shorter, more accessible 1952 version is the one still taught in schools.

No one can determine what Anne's final draft might have been like, but to ignore the time and energy she put into version "b" is to deny her own ideas about what she wanted her book to be, insofar as we can know them. Laureen Nussbaum makes a case for valuing Anne's literary judgment: "A reader poring over the *b* version will find it hard not to look at the parallel printed *a* version in order to make comparisons. In doing so, *this* reader could not help but be impressed with the amount of self-criticism and literary insight the barely fifteen-year-old Anne brought to bear upon her revision, omitting whole sections, reshuffling others, and adding supplementary information so as to create a more interesting and readable text. In the process, she must have used all her writing talent and the know-how gleaned from her extensive reading . . .

"My conclusion: readers who appreciate a well-written book, but who are not necessarily into women's studies or literary criticism, have a right to read Frank's wartime story in a form as close as possible to the author's own final version. Conversely, we owe it to Anne Frank that at long last she be taken seriously as the writer she really was, before the Disney people market her as their next popular heroine, Pocahontas-style."

MONTHS before the minister-in-exile's radio speech inspired Anne to go back and begin her diary again, she wrote an entry that explains why she took so readily to the project of massive revision. In the midst of a paper shortage, during her second winter in hiding, she searched the diary for pages that had been left blank, and filled them in. On one such page, dated January 22, 1944, Anne describes the shock of confronting the writing of a younger self. If Philip Roth noted that reading the diary was *like watching a fetus grow a face*, the face that has grown by this point is that of an author realizing that her early work could be improved upon.

> When I look over my diary today, 1 ½ years on, I cannot believe that I was ever such an innocent young thing. . . . I still understand those moods, those remarks about Margot, Mummy and Daddy so well that I might have written them yesterday, but I no longer understand how I could write so freely about other things. I really blush with shame when I

read the pages dealing with subjects that I'd much better have left to the imagination. . . . This diary is of great value to me, because it has become a book of memoirs in many places, but on a good many pages I could certainly put "past and done with."

PART III

The Afterlife

SIX

The House

THE ANNE FRANK MUSEUM OPENS AT NINE EVERY morning, and by ten, even on cold winter days, a line reaches the corner, straggling along the sidewalk across from the picturesque Prinsengracht Canal. The majority of the people in line are young, as are the majority of visitors to Anne Frank's house. Quite a few of the adults seem subdued, uneasy, perhaps because of what they're about to encounter. But though the high school students understand that this is not supposed to be fun, their investment in coolness dictates that they exude the detached nonchalance of kids about to be taken through any art gallery or royal palace, or a guild hall where some historic treaty was signed.

Inside, nothing about the cheerful, modern, brightly lit reception area suggests that this will be any different from any other museum experience. Credit cards and cash are surrendered, tickets issued. But as soon as one enters the house itself, even the most garrulous teens fall silent, and a hush falls over the visitors. It's hard to walk through the former offices of the Opekta staff, then up into the storeroom and past the bookcase that once concealed the door to the secret annex without bordering on, or crossing over, the edge of tears.

Part of what makes the Anne Frank Museum so affecting is its simplicity and the sense that very little has changed since the furniture movers stripped the premises bare in the wake of the Jews' arrest. Here and there, a video monitor plays an informational film strip—one about the role of the helpers who aided the Franks, another featuring an interview in which Hanneli Pick-Goslar describes her last sight of Anne at Bergen-Belsen. Here and there, simple glass vitrines display a few of the objects that remain from that period: Miep Gies's identity card, Edith Frank's prayer book. Here and there, a quote—from Primo Levi, or from Anne's diary—has been stenciled on a wall. There are photos of the eight people who hid here, another of Jews being rounded up on an Amsterdam street. But these mostly bare rooms, these walls and floors and ceilings, are allowed to speak for themselves. One never feels that strong emotions are being artificially manufactured and extracted; nowhere is there a trace of the sentimentality and kitsch so problematic in, say, the Berlin Holocaust Memorial, where visitors must trudge through a gallery partly filled with clanking metal frowny faces intended to represent the murdered Jews of Europe.

The decision to keep the rooms unfurnished was made by Otto Frank, who felt that the secret annex should appear just as it did after his family and all their possessions were seized. But in fact a great deal of labor, planning, construction, and restoration has been required to give visitors the impression that nothing has been touched.

In the early 1950s, a textile company, Berghaus, bought the block on which the Opekta company stood, and announced its plans to raze the old houses and businesses and replace them with a modern office. By then, the diary had already acquired an international reputation, and its fans had begun making pilgrimages to 263 Prinsengracht, where Johannes Kleiman and others often agreed to take them on informal tours. With the support of Amsterdam's mayor, a campaign was initiated to prevent the proposed demolition. Faced by widespread opposition, Berghaus withdrew, and a fund-raising drive gathered the requisite capital to buy the property. The success of this drive enabled the establishment of the Anne Frank Foundation, in May 1957, and Otto Frank financed the purchase of the building next door, specifying that it be used as an educational center.

Three years later, the Anne Frank Museum was officially opened. By then, thanks partly to the play and the film based on the diary, the book's popularity had increased exponentially, and in the first year of operation, the museum hosted 9,000 visitors. In 1970, 180,000 people came to see Anne's secret annex, by which point the volume of foot traffic necessitated architectural improvements to shore up and maintain the site. The front part of the building—the former Opekta office—was modernized to include a reception center, while the annex was left as it was.

As the number of visitors grew each year, an expansion plan evolved, and in the 1990s, a new building was added onto the museum. In contrast to the labyrinthine, confined spaces of the secret annex, the new structure that houses the reception area, the bookstore, and the cafeteria is airy and expansive. At the same time, the Opekta office was restored (with the aid of old photographs and floor plans, and with meticulous attention to the period authenticity of every doorknob and light switch) to allow museumgoers to experience something of the atmosphere of the rooms in which Miep Gies and her colleagues continued, throughout the war, to operate the business that sustained the Jews. More recently, in 2008, the scale model of the secret annex that Otto had made was given a permanent place, and the room in which the checked diary is kept was redesigned to emphasize the central importance of Anne's book.

But for the hundreds of school groups and the million visitors who, in 2007, passed the movable bookcase and climbed the steep stairs to the top of the building, the heart and soul of their visit is Anne's room. Hardly anyone speaks as they file past the postcards that Anne glued on her wall: photos of cockatoos and wild strawberries, of movie stars, of the British royal princesses, and of chimpanzees at a tea party. Once more, the unspoiled simplicity of the room is eloquently communicative of everything that transpired inside it. The main difference is that, when the Franks and the others were in hiding, the windows, now admitting the bright Amsterdam sun, were covered, for safety. Only from the garret, which it is no longer possible to enter but which can be glimpsed in a mirror, could Anne see the sky outside, or the night stars, or the blossoms in her beloved chestnut tree, announcing that another spring had come.

AROUND the corner from, and attached to, the Amsterdam warehouse above which the Franks spent twenty-five months, the Anne Frank Foundation is, in effect, the annex to Anne's secret annex. Its corridors are decorated with posters, framed book covers, and images documenting the diary's impact. On one wall is a photo of Nelson Mandela, who found encouragement in Anne Frank's account of her incarceration in an attic during his own long imprisonment on Robben Island. Tacked to a bulletin board is a snapshot of two Afghan girls holding the diary translated into Dari, while a small shelf has become a sort of folk-art shrine featuring a portrait of Anne painted by a Russian student, another stitched in needlepoint by a reader from Ukraine.

More than fifty years after it was established, the foundation now has over a hundred employees. Staff members direct the day-to-day operation of the Anne Frank Museum and raise funds to preserve the warehouse and the attic, where the wallpaper must be regularly replaced because so many pilgrims cannot resist the urge to touch it. The foundation develops educational materials, supports a traveling exhibition, monitors incidents of racism worldwide, facilitates research about the Holocaust and human rights, and oversees the Anne Frank archive. It publishes a quarterly journal and maintains a Web site, which, in 2006, attracted three million visitors.

It also operates in cooperation with the Anne Frank-Fonds in Basel, a separate institution—headed by Anne's only surviving relative, her cousin Bernd Elias—that controls the rights to Anne's writings and oversees humanitarian projects, including a school for India's untouchables, a Jewish-Arab youth orchestra in Jerusalem, and a program to train teachers to work among the poorest citizens of Peru. On occasion, the two organizations have come into conflict. In 2008, the Anne Frank-Fonds questioned the appropriateness of a musical based on the diary and staged in Madrid, while the Anne Frank Foundation in Amsterdam, where the play's authors and stars had come to do research, viewed the production as a means of introducing Anne to new audiences. But by and large, the two institutions have worked amicably to facilitate the delicate balancing act of both publicizing and protecting Anne's image, and her book.

In a sunny office on the third floor of the Anne Frank Foundation, a young Argentinian woman named Mariela Chyrikins sorts through a stack of photographs. Taken in Argentina, in 2006, the pictures show a group of police cadets whose full-dress uniforms make them appear to be standing at attention even when they're relaxed. Behind them are partitions and panels displaying an exhibition about the life and legacy of Anne Frank, whose diary they have gathered to discuss.

As in most of the 177 sites that hosted "Anne Frank—A History for Today" in 2006, venues ranging from Slovakia to Reno, from Croatia to Cyprus, the Argentinian program, held in Cordoba and Buenos Aires, involved local volunteers—in this case, chosen from among the cadets. These volunteers were trained to serve as guides who would teach their peers about Anne Frank and the Holocaust, and lead discussions that, it was hoped, would inspire the participants to talk about their own experiences with bigotry and violence. During the workshop, the cadets met for lunch with a delegation of university students—two groups that normally would never sit down together, let alone speak about what had happened to their families and their communities during the brutal military dictatorship that ruled Argentina from 1976 to 1983.

As a teenager, Mariela Chyrikins was drawn to the diary in the shocked aftermath of the 1994 bombing of the AMIA Jewish Community Center, which killed eighty-five people in her Buenos Aires neighborhood. She began to write, then to e-mail, the Anne Frank Foundation. She helped bring the Anne Frank exhibition to Argentina, then followed it back to Amsterdam, where, according to Jan Erik Dubbelman, head of the foundation's international department, "she made us realize we had no choice but to hire her."

"The moment that young people in Latin America become emotionally involved with Anne Frank's story," says Mariela, "they start to reflect about themselves. They say, 'Ah, okay, in Europe there also happened something like what happened in our country.' It's like opening a Pandora's box, you never know the consequences of the work you are doing, but you know that you are touching people's hearts.

"The curriculum at the police academies is left over from the times of the dictatorship, very authoritarian. And when they put on the uniform they have the feeling that they are no longer a human, but a god. Anne Frank's story makes them more human. The cadets want to protect Anne Frank, as if she was their own child, as if she's their best friend. They're mad because Anne Frank was killed, but when they see some poor guy working on the street, they might want to kill *him*. The important thing is making a connection between this poor guy and Anne Frank.

"Maybe their parents were also police during the dictatorship, but they have never talked about it. Now they say they are going to talk to their parents. They have lunch with the university students, and for the first time they start to discuss: Who is the other, how do you see the other? They are human beings too! In some cases, the children and grandchildren of the disappeared are among the trainers. It brings a lot of very emotional things to the surface. It was hard for the police cadets to listen to the stories of the Dirty War. But Anne Frank's story allows them to be inside someone else's shoes. It allows young people in Latin America to raise their voices—and to flower."

Similar programs have taken place in Guatemala and in Chile, at the Villa Grimaldi, a former torture center under the Pinochet regime and currently a memorial to its victims. In both countries, Anne's diary has enabled its readers to confront their troubled past—and, in Guatemala, to discuss the ongoing violence that is still part of daily life.

Down the hall from Mariela Chyrikins's office, Norbert Hinterleitner is engaged in a similar project, one that involves bringing programs about tolerance to provincial cities in Ukraine. Born and raised in Austria, Norbert has spent the past five years designing books and textbooks, distributing copies of Anne's diary, and working with educators and teenagers to combat not only anti-Semitism but xenophobia, homophobia, and bias against the region's ethnic and racial minorities. It's tricky, Norbert admits, even to gain admittance to schools and community centers whose administrators are more likely to hang up on him than to answer his phone calls. On an early visit to the former Soviet Union, he used, as a teaching tool, a recent article from a local newspaper, a four-page "factual" exposé of the worldwide conspiracy masterminded by American Jews in secret partnership with the Mafia.

Norbert says, "If you're surprised and you show it, you've lost." He's satisfied if he can reach a few kids among the thousands, kids who normally never speak up, kids whose lives will be changed if they can be persuaded that they are not alone and that it is all right to take a public stand against bigotry. What makes Anne's diary so useful, he has found, is Anne's fundamental decency, her belief that human dignity will prevail.

"She was a victim of her society, but when you talk about her book, it gives people hope and inspiration. It's a catalyst. They begin to think that they can do something different." Norbert's sincerity is infectious, but I can't quite stifle the skeptical thought that, given the persecutions and pogroms that have transpired in that region of Eastern Europe, teaching certain Ukrainians not to be anti-Semitic is a bit like trying to teach cocker spaniels to fly. What could the success rate of such a program be, and how could it be quantified? Norbert, I think, is probably wise to keep his ambitions modest, to attempt to reach a few receptive kids. But despite my reservations, I'm delighted by Norbert's hopeful determination. I'd like to imagine that he is right in his faith that entrenched hatred can be dissolved and dispelled, one Ukrainian teenager at a time.

True, the ending happens just as the Franks and their friends had feared all along: their hiding place is discovered,

*and they are carried away to their doom. But the fictitious
declaration of faith in the goodness of all men which concludes the play falsely reassures us since it impresses on us
that in the combat between Nazi terror and continuance of
intimate family living the latter wins out, since Anne has
the last word. This is simply contrary to fact, because it was
she who got killed. Her seeming survival through her moving
statement about the goodness of men releases us effectively of
the need to cope with the problems Auschwitz presents . . . It
explains why millions love the play and movie, because while
it confronts us with the fact that Auschwitz existed it encourages us at the same time to ignore its implications. If all men
are good at heart, there never really was an Auschwitz; nor is
there any possibility that it may recur.*
—BRUNO BETTELHEIM, *"The Ignored Lesson of Anne Frank"*

On the morning after my first visit to the Anne Frank Foundation, I cancel my appointments and spend the day in my hotel
room so I can process what I've seen and heard. It's not only that
I've been moved by Mariela's energized idealism, by Norbert's
sweet, quixotic hope that the ugly story of Eastern European
anti-Semitism can have a different ending. It's also that my conversations with them have changed my way of thinking about
Anne Frank's diary, and about the ways in which it has been
received.

I had become increasingly impatient with the notion of
Anne Frank as the perky teenage messenger of peace and love,
as a source of what Ian Buruma has termed "kitsch absolu-
tion," a modern-day saint preaching tolerance from beyond the
grave—in this case, a mass grave at Bergen-Belsen. Such a mis-
reading of Anne's book and of her "message," I'd thought, con-
stituted a denial of what happened to her after the diary ended,
and of the cruel fates that befell millions of equally innocent

men and women and children. That is what Bruno Bettelheim concludes in the paragraph above, taken from an otherwise quite mad essay in which he blames the Franks for the arrogance of insisting on going into hiding as a family, as well as for the crime of not having survived.

The emphasis on redemption and forgiveness seemed all too reminiscent of the saccharine endings of the Broadway drama and the Hollywood film based on the diary. The play ends with Anne's statement of her belief that, despite everything, people are really good at heart. At the conclusion of the film, music soars, birds twitter, the camera ascends toward the puffy clouds dotting the calm sky, while, on the sound track, the girlish fashion model playing Anne Frank reaffirms her faith in humanity. Clearly, people, or *some* people, are good at heart, but the reality of Anne's story, the reality of Auschwitz and Bergen-Belsen, would suggest that *some* people are basically evil at heart. "The line that concludes her play," wrote Holocaust scholar Lawrence Langer, "floating over the audience like a benediction assuring grace after momentary gloom, is the least appropriate epitaph conceivable for the millions of victims and thousands of survivors of Nazi genocide."

In fact, Anne herself had a sensibly and understandably mixed view of human nature. Among the most impressive aspects of her diary is the way in which its author is able to entertain and even embrace two apparently irreconcilable ideas about mankind. Anne's book is a testament to certain individuals' ability to develop, at an early age, a sophisticated moral consciousness, and to maintain compassion and humor under the most intense stress. Her "ambivalence about the hard questions of life" was, Buruma noted, a "mark of her intelligence."

On May 3, 1944, Anne wrote, "I don't believe that the big men, the politicians and the capitalists alone, are guilty of the war. Oh, no, the little man is just as guilty, otherwise the peoples of the world would have risen in revolt long ago! There's in people simply an urge to destroy, an urge to kill, to murder and rage, and until all mankind, without exception, undergoes a great change, wars will be waged, everything that has been built up, cultivated, and grown will be destroyed and disfigured, after which mankind will have to begin all over again."

Almost three months later, and two weeks before her arrest, Anne composed the entry with which, for better or worse, she would become most closely identified. It's worth quoting the entire passage as a corrective to the simplistic, falsely consoling idea of Anne Frank as an endlessly optimistic spirit. What's striking is how beautifully written the entry is and how, like its author, it veers between extremes of hope and despair.

> *Anyone who claims that the older ones have a more difficult time here certainly doesn't realize to what extent our problems weigh down on us, problems for which we are probably much too young, but which thrust themselves upon us continually, until, after a long time, we think we've found a solution, but the solution doesn't seem able to resist the facts which reduce it to nothing again. That's the difficulty in these times: ideals, dreams, and cherished hopes rise within us, only to meet the horrible truth and be shattered.*
>
> *It's really a wonder that I haven't dropped all my ideals, because they seem so absurd and impossible to carry out. Yet I keep them, because in spite of everything I still believe that people are really good at heart. I simply can't build up my hopes on a foundation consisting of confusion, misery, and death. I see the world gradually being turned into a wilderness, I hear the ever approaching thunder, which will destroy us too, I can feel the sufferings of millions, and yet, if I look up into the heavens, I think it will all come right, that this cru-*

*elty too will end, and that peace and tranquillity will return
again.*

The sentence about human goodness has, as Cynthia Ozick
observed in a searing 1997 *New Yorker* piece, "been torn out
of its bed of thorns." In her essay "Who Owns Anne Frank?"
Ozick raged at how easy and how common it has become to dissociate the story of Anne Frank's life from its tragic conclusion;
playwrights, producers, and publishers have "bowdlerized, distorted, transmuted, traduced, reduced . . . infantilized, Americanized, homogenized, sentimentalized; falsified, kitschified,
and, in fact, blatantly and arrogantly denied" the truth of the
diary. That truth, claimed Ozick, resides in the crimes listed in
the Nazi transport lists, which record that Anne Frank and the
others were deported to Auschwitz on September 3, 1944, along
with 1,019 "*Stucke* (or 'pieces,' another commodities term)."

By now, we know how Anne Frank's story ended. The eye-
witnesses have been deposed. We know about the horror of her
imprisonment in Auschwitz and about her death, from typhus
and malnutrition, at Bergen-Belsen.

The fact that Anne's journal concludes before the Franks'
arrest and deportation has been viewed as one of its short-
comings. In *Literature, Persecution, Extermination*, Sem Dresden
argues that Anne's diary is "not gruesome enough." "The diary
is taken to be a Holocaust document," wrote Cynthia Ozick.
"That is overridingly what it is not."

In his essay in the *New Republic*, Robert Alter agreed. "A
girl's journal about her family in hiding that ends with an editor's note about her fate cannot convey the full actuality or
meaning of a catastrophe in which millions of individuals and
much of their culture were obliterated in camps built and oper-
ated by one of the great nations of Europe." *Ends with an edi-*

tor's note implies that Anne Frank's diary is incomplete, as if she missed the main event, which, as we know, she did not. "There are no skeletal camp inmates, no gas chambers, no diabolical medical experiments and acts of sadism."

But according to Norbert Hinterleitner, one of the reasons the diary is such a useful teaching tool is *because* it allows students to develop an attachment to Anne Frank before they learn about the horrors of the Nazi camps. "It's full of fear, but not of suffering." The semblance of ordinary domesticity that the Franks preserved enables Anne's audience to read her story without feeling the desire to turn away, the impulse we may experience when we see the photos and footage of the skeletal dead and dying.

It has also been argued that there is something false or at least distorted about viewing this utterly singular girl as representative of the millions who were murdered. But whether we approve or not, her individuality is the reason—in some cases the only reason—students everywhere are taught about the Holocaust. "Statistics don't bleed," wrote Arthur Koestler, "it is the detail which counts." A similar notion is expressed in the quote from Primo Levi that appears on a wall of the Anne Frank House Museum: "One single Anne Frank moves us more than the countless others who suffered just as she did, but whose faces have remained in the shadows. Perhaps it is better that way: If we were capable of taking in the suffering of all those people, we would not be able to live."

Reviewing the diary in the July 1952 *Saturday Review*, Ludwig Lewisohn wrote, "If Anne Frank's diary pierces the conscience of men in all its implications—the implications of her being and her people's being, of her life and of her death . . . if that were possible, the publication of her diary would indeed be a moral event of inestimable import. . . . A million Anne

Franks died in horror and misery; millions of human souls are perishing similarly today in the Soviet Union, in the 'People's Democracies' in China. . . . Contrition for Anne Frank may rouse other contritions and from this one girl's diary a gleam of redemption may arise."

Perhaps some part of that contrition *does* depend on our catching our last sight of Anne before she was stripped of everything we associate with being human. At the same time, it is crucial that the diary be read in its historical context, just as it is a distortion of everything Anne suffered to treat her book as the story of a teenager's problems with her mother. It is essential to point out that, although there have been, and likely will be, other genocides, the methodical efficiency with which the country that had produced Goethe and Bach set out to eradicate an entire population is still so far unparalleled. Once the historical background *has* been established, once it is made clear that Anne's being forced into hiding and murdered was the direct result of the Nazis' plan to exterminate the Jews, why *not* identify with Anne Frank as a suffering fellow creature? As Ian Buruma observed, "Such identification can result in sentimental self-pity, but it is more likely to give people at least some idea of the evil that was done."

In "Who Owns Anne Frank?" Cynthia Ozick railed against the American teenagers who burdened Otto Frank with the adolescent angst they imagined to be just like his daughter's. Ozick derided Otto's compulsion to answer their goofy, heartfelt letters, and singled out, as an especially egregious example, the 1995 book *Love, Otto*, drawn from the long correspondence between Otto Frank and a California girl who auditioned for the role of Anne in the Hollywood film. In one letter, Cara Weiss, later Cara Wilson, wrote, "Despite the monumental difference in our situations, to this day I feel that Anne helped me through the teens with a sense of inner focus. She spoke for me. She was strong for me. She had so much hope when I was ready to call it quits." Ozick concludes, "The unabashed triflings of Cara Wilson—whose 'identification' with Anne Frank can be duplicated by the thousand, though she may be more audacious than most—point to a conundrum. . . . Did Otto Frank not comprehend that Cara Wilson was deaf to everything the loss of his daughter represented? Did he not see, in Wilson's letters alone, how a denatured approach to the diary might serve to promote amnesia of what was rapidly turning into history?"

But, one might ask, why *not* encourage and guide young readers' responses? Why should they *not* imagine they have something in common with a girl whose life, so unlike their own protected ones, ended in Bergen-Belsen? Why should the diary not inspire people like Mariela Chyrikins and Norbert Hinterleitner to attempt to rid the world of hatred? Why *not* emphasize Anne's optimism if it means that one Argentinian police cadet or one Ukrainian high school student might be more responsive to the humanity of others?

The Nazis understood how useful it was to prevent the camp guards from identifying with the prisoners, to emphasize the otherness, the difference of the people whom the boxcars brought to Sobibor and Treblinka. In Gitta Serenyi's book, *Into That Darkness*, Franz Stangl, commandant of the Treblinka extermination camp, explained that the brutality of the camp routine—the whips, the shouting, the stripping of the new arrivals and the forcing of them to run—was devised not for its effect on the prisoners but rather for the benefit of the guards, who would work more efficiently to the degree that they believed their victims were not human beings.

In my Amsterdam hotel room, I have what I suppose could be called a change of heart. It no longer strikes me as quite so reductive to see Anne's "message," as Otto Frank wished it to be interpreted, as one of tolerance and understanding. Poor Otto!

As if losing his wife and daughters were not enough. How he has been reviled for believing that Anne's diary could leave its readers more conscious and compassionate! How often he has been accused of purging the book of Judaism, of sex, of its ferocious mother-daughter conflict, when in fact all those elements exist in the version he edited. And how he has been vilified for allegedly having turned (in Ozick's phrase) a "deeply truth-telling work" into "an instrument of partial truth, surrogate truth, or antitruth," and for having given his blessing to a foundation that has "washed away into do-gooder abstraction the explicit urge to rage that had devoured his daughter."

It has yet to be explained to me how the existence of a human rights foundation denies and negates the sufferings of Anne Frank and so many others, how viewing her life exclusively through the lens of her death will bring her back from the dead, or how it will forestall a recurrence of the virulent anti-Semitism responsible for her murder. If it turns out to be even remotely possible to prevent further horrors, surely Mariela and Norbert are doing more to achieve that end than those who mock them as naive do-gooders? Once again, I am struck by the uniqueness—the anomalousness—of Anne Frank's diary, which still moves people to fierce debate, to talk about the political and social realities of the society they inhabit, and about the life of a girl who hid for two years in an attic until she was arrested and deported and killed.

THERE is, in the library of masterpieces, an entire subcategory of books whose authors could be said to have been forced into a collaboration with misfortune. Among the depressingly numerous products of that involuntary partnership are *Hope Against Hope*, Nadezhda Mandelstam's account of the terror of life under Stalin, and *Into the Whirlwind*, Eugenia Ginsburg's

memoir of that same period. There are the poems of Szymbor-
ska, Milosz, and Celan, Primo Levi's *Survival in Auschwitz*, the
slave narratives of the antebellum South, all manner of war and
prison novels.

These are books that came into being at a personal cost that
no one would be willing to pay. Their authors had no choice but
to endure the circumstances that led to their books' composition, and the books were what they had to show for it, if they
survived. It is likely that none of them would have written their
novels and poems and memoirs if they could have avoided their
subjects, if their subjects had not sought them out, or hunted
them down. All of which makes it problematic for us to say how
good the books are, and how grateful we are that they exist.

Given the choice, we would have been willing to live without the diary if it had meant that neither Anne Frank nor anyone
like her, or anyone *unlike* her, had been driven into hiding and
murdered. But none of us was given that choice, and the diary
is what we have left. Meanwhile, across the equator and around
the world, Anne Frank's strong and unique and beautiful voice
is still being heard by readers who may someday be called upon
to decide between cruelty and compassion. Guided by a conscience awakened by a girl in an Amsterdam attic, one citizen
of Ukraine or one Argentinian policeman may yet opt for humanity and choose life over death.

SEVEN

The Play

THE SAGA OF THE BROADWAY PRODUCTION OF *THE DIARY of Anne Frank* is so rife with betrayal and bad behavior, so mired in misunderstanding and complication that at least four books have attempted to explain what happened and why. Published in 1973, Meyer Levin's aptly titled *The Obsession* blames a left-wing cabal masterminded by Lillian Hellman, a secret conspiracy to purge the diary of everything Jewish, including the six million dead. Levin's version is supported by Ralph Melnick's *The Stolen Legacy of Anne Frank*, which likewise sees Hellman as the malevolent puppeteer in the wings.

Ten years after Meyer Levin's death, his wife, Tereska Torres, wrote a book, *Les Maisons hantées de Meyer Levin*, which appeared in France and mingled fiction and memoir to portray a long and loving marriage to a man whose every hour was haunted by Anne Frank's ghost. The most dispassionate analysis of the controversy can be found in Lawrence Graver's *An Obsession with Anne Frank*. But though these *Rashomon*-like retellings of the drama surrounding the drama disagree on the motives of the principal players and the machinations of the supporting actors, all are more or less in accord about the basics of the scenario, in which high-mindedness and slipperiness coexisted in extremely elevated concentrations, and which spanned decades of accusations and counteraccusations, decades in which money, power, and fame were pursued in the name of idealism and of loyalty to a murdered girl.

Eventually, the unfortunate history of the dramatization of Anne Frank's diary would become as convoluted as the plot of a Dickens novel. Cynthia Ozick compared the story to Jarndyce versus Jarndyce, the protracted court case at the center of *Bleak House*, and indeed the wrangling, the maneuvering, the charges and countercharges would recall not only the obsessional lawsuit that Dickens so brilliantly portrayed, but also his awareness of what such a fixation costs, even beyond the legal fees.

WITH eight major characters, a few rooms, one set, a rising arc of family conflict, teenage romance, and terror, Anne Frank's diary seemed perfect for the stage. Within days of the American publication of *The Diary of a Young Girl*, Doubleday's New York office was fielding calls from interested producers. The callers must have assumed that they were getting in at the beginning. But as those who became involved in the production would soon discover, to their chagrin, these early arrivals had actually come to the story late, seven years after it started in the same place where Anne Frank's story ended, and only a few months after her death.

A prologue of sorts had been enacted over a mass grave at Bergen-Belsen, to which an American writer named Meyer Levin traveled as a correspondent for the Overseas News Agency and the Jewish Telegraphic Agency. A scrappy Chicago tailor's son, the child of Eastern European immigrants who had risen, through hard work, to upper-middle-class respectability, Levin had had a gnarled relationship with his Jewish heritage. But those tangles sorted themselves out when Levin witnessed the Allied liberation of the Nazi prison camps.

By then, he had published several well-received novels, none a commercial success, and had been cobbling together a living by writing journalism, criticism, and short fiction. Too old for active service, he was glad to have found a way to use his skills in the war effort.

His instincts in the face of catastrophe were generous and energetic. He was known for asking newly freed prisoners if there was anyone they wanted him to contact. Survivors inscribed their names in the dust on his Jeep. Levin vowed to make sure America knew about the destruction of the European Jews, and though he briefly considered writing about their fate, he became convinced that "from amongst themselves, a teller must arise."

After the war he helped Jewish refugees trying to reach Palestine, made two documentary films about their plight, and wrote *The Search*, a three-part memoir about his Jewish-American childhood, what he observed in the camps, and his work to help survivors emigrate to Israel. The book was rejected by every editor to whom Levin showed it, one of whom criticized its excessive whining about anti-Semitism. It was finally printed in Europe at Levin's expense and by a small publisher in the United States.

In 1950, Levin and his wife and their two children moved to the south of France so he could work on a screenplay adaptation of an early novel about a musician—a project commissioned by the violinist Yehudi Menuhin, who wanted to make a movie in Israel. In Antibes, Tereska Torres gave Levin a copy of Anne

Frank's diary, which had just appeared in French. It was a gift that Tereska, a writer whose Polish father had converted from Judaism to Catholicism, would regret.

As he read the diary, Levin became convinced that Anne Frank's was the voice he had prayed to hear as he stood over the mass grave at Bergen-Belsen. He felt the excitement familiar to anyone who has ever discovered an unknown masterpiece and been convinced of the importance—the *necessity*—of its reaching a wider audience. But in fact the book was hardly unknown, having made the rounds of American and British publishers and been everywhere rejected.

When Levin wrote Otto Frank, praising the diary and offering to help find it a home in the United States, his letter must have seemed to Otto like a reason for fresh hope. Levin reassured Otto that his enthusiasm for the book had nothing to do with money. He himself would translate the diary if Otto thought it might be useful. He mentioned an obvious selling point, the story's potential as a play or even a film. Otto Frank was less convinced about the diary's dramatic possibilities, but he gratefully accepted Levin's offer to broker its sale.

After the first round of letters, Otto Frank and Meyer Levin became friends. Their correspondence suggests an exchange between a fond uncle and his younger, smarter, savvier—but always respectful—nephew. Promising to make the right contacts and to help Otto navigate the treacherous currents of American publishing without sacrificing the integrity of Anne's work, Levin began, with Otto's blessing, a letter-writing campaign to American editors.

In November 1950, after the *New Yorker* ran Janet Flanner's mention of the diary's French success, Little, Brown offered to copublish the book with Vallentine-Mitchell in the UK, but the offer fell through when Little, Brown insisted on retaining the dramatic rights. Persuaded, presumably, by Levin, Otto had become so convinced of the necessity of controlling these rights that it was the one provision he insisted on in his negotiations with Doubleday, which acquired the book when Little, Brown dropped out.

Judith Jones's account of finding the French translation of Anne Frank's diary in the Doubleday rejection pile and of reading all afternoon and into the evening contains a brief coda describing a conversation that took place after Jones convinced her boss, Frank Price, to publish the diary. When Otto Frank asked to meet with representatives from Doubleday's European office, Price and Jones invited him to come from Amsterdam to Paris. After a long, leisurely lunch, "he made just one stipulation. He wanted to have a say in the dramatic rights, because he admitted, with tears in his eyes, 'I couldn't bear the thought of some actress playing my Annie.'" Time would show that Otto's misgivings were correct, for more reasons than he could have imagined.

In the early letters between Otto and his editors at Doubleday, Meyer Levin initially appears as a beneficent presence who wanted the best for the diary and who was to be consulted on decisions about the American edition. Levin's good intentions became even more apparent, and his intercession more welcome, when his *Times* review propelled the diary onto the best-seller lists.

Levin was not responsible, as he would claim, for the diary's publication, but he was important in its success, and that success was his undoing.

THE Obsession is a strange memoir, a journal of madness by a madman who has yet to be cured, or for that matter convinced that the pathology that has ruined his life is an illness. It begins with a nod to Dante's descent into hell: "In the middle of life I fell into a trouble that was to grip, occupy, haunt, and all but devour me, these twenty years." The first scene takes place in the office of the latest of several psychiatrists to whom Levin has gone in search of insight, if not relief. "Amazing, how these writers carry on, running from one analyst to another, the way, after a pessimistic medical diagnosis, one runs to another specialist in search of a different finding!"

Digressive, discoursing on the evils of McCarthyism and the reluctance of the Soviet Union to let its Jewish citizens leave for Israel, shot through with complaints about the literary establishment's conspiracy to ignore his latest novel, *The Obsession* keeps looping back to its central theme in a way that will be familiar to anyone who has ever tried to talk to the severely depressed. Comparing himself to Solzhenitsyn, a fellow victim of political persecution, Levin exhibits delusions that he never suspects are delusional, any more than he realizes that the opposition lawyer whom he mocks for saying, "Levin has the hallucination that he actually wrote the diary," could hardly have been more correct.

"Agreed, it was an obsession. Admitted. There it sat under my skull with my mind gripped in its tentacles. Sometimes dormant. Sometimes awakening and squeezing. Again I would react, send out protests and petitions. That was all very well for suppressed Russian writers, from prisons, from labor camps . . . but for a free American writer to complain for twenty years about a so-called act of suppression was obviously obsessional."

The Obsession is suffused with Levin's intense and unconscious ambivalence. He describes Holocaust survivors as "ringed by eternal fire in the unapproachable area of those who have endured an experience that puts them beyond our judgement" and on the very next page reports telling Otto Frank, "You have been my Hitler." His wife appears to have been a paragon of patience and forbearance until Meyer's fixation drove her to attempt suicide and nearly wrecked their marriage. "Again and again times have come when she agonizedly cried out, 'It's me or Anne Frank! Choose!' as though this were some rival love I could abandon at will (Masochist, clinging to your pain-giver!). But can one by an act of will banish what invades one's mind?"

Levin's claim to have arranged the diary's popularity might seem like a wishful fantasy if not for the front-page rave in the *New York Times Book Review*. We can assume that his review was motivated by unalloyed admiration for a book that went on to be admired by millions, and that he meant his praise wholeheartedly. Yet the cheapening of Anne Frank's diary had already begun, set in motion, with the best intentions, by a man who dedicated himself to ensuring that it not be cheapened.

"Anne Frank's diary," Levin's review begins, "is too tenderly intimate a book to be frozen with the label 'classic,' and yet no lesser designation serves. For little Anne Frank, spirited, moody, witty, self-doubting, succeeded in communicating in virtually perfect, or classic, form the drama of puberty. But her book is not a classic to be left on the library shelf. It is a warm and stirring confession, to be read over and over for insight and enjoyment.

"The diary is a classic on another level, too. It happened that during the two years that mark the most extraordinary changes in a girl's life, Anne Frank was hidden with seven other people in a secret nest of rooms . . . The diary tells us the life of a group of Jews waiting in fear of being taken by the Nazis. It is, in reality, the kind of document that John Hersey invented for *The Wall*."

Already, the essay has become a kind of pitch, with the pitch's nod to the latest work on a similar theme that made money. Ironically, Levin had written an article in *Congress Weekly* arguing that Hersey's best-seller about the Warsaw ghetto had found an audience that would be denied Anne's diary and Levin's *In Search* because the authors were Jews. In the *New York Times Book Review,* Levin points out that the diary "probes far deeper into the core of human relations, and succeeds better than *The Wall* in bringing us an understanding of life under threat." Which is true. Today *The Wall* is hardly read. But for Levin, it was the competition.

Yet another striking feature of the *Times* review is Levin's use, before the first space break, of the word *universalities*—a term he came to despise as he argued for the particularity of Anne's experience. "It has its share of disgust, its moments of hatred, but it is so wondrously alive, so near, that one feels overwhelmingly the universalities of human nature. These people might be living next door; their within-the-family emotions, their tensions and satisfactions are those of human character and growth, anywhere." Levin's emphasis on the people-next-doorness of the hidden Jews would later be echoed by all those who wanted the book, the play, and the film to have the largest possible audience. *Universal* is not just an adjective, but, in the world of commerce, a projected number, which is why *universal* would be employed, more and more frequently, as the antonym of *Jewish.*

THE *New York Times* was understandably upset by Levin's failure to inform them of his connection to the diary. In the Anne Frank Museum archive is a letter from Meyer Levin to the *Book Review,* expressing his hope that the paper will ask him to write for them again. He understands that the editors have heard he was agenting Anne Frank's diary, which might suggest that it would have been unethical of him to request the review assignment. But in fact he wasn't, strictly speaking, the agent. He had no intention of *profiting* from the book and had been motivated by pure enthusiasm for the diary as literature. Later, Levin would blame Barbara Zimmerman for suggesting he seek the assignment from the *Times*.

No sensible person would argue that books should be reviewed by their agents. But *The Diary of a Young Girl* may be an anomalous case that encourages us to view even this dubious event in light of its result. What if the diary had been assigned to a critic who, like the reader at Knopf, thought it "a dreary record of typical family bickering, petty annoyances and adolescent emotions"? Even though Eleanor Roosevelt had praised the book, a lukewarm review might have arranged a slow trip to the remainder table, where, with luck, it might someday be rediscovered for the Holocaust Studies series of a university press. Arguably, Levin's review represented a breach of ethics that worked out for the best.

Should Doubleday have told the *Times* not to run the review because Levin wanted to write the play? The publisher believed in the book; it would have been self-defeating. The letters between Otto Frank and Barbara Zimmerman exulted over Levin's essay, which Zimmerman praised for its beauty and for the space it got in the country's most influential paper. Buoyed by the book's success, everyone chose to ignore the fact that Levin was in touch with at least two Hollywood studios and that *Variety* described him as "agenting the tome for possible filmization or legit treatment."

Though Levin pitched the diary as being better than *The Wall*, a successful pitch needs a successful precedent, and in this case there was none. There had been nothing quite like it, and no evidence suggested that adult readers would buy a girl's diary. Meyer Levin argued that the book should be read: "There is anguish in the thought of how much creative power, how much sheer beauty of living, was cut off through genocide. But through her diary Anne goes on living. From Holland to France, to Italy, Spain. The Germans too have published her book. And now she comes to America. Surely she will be widely loved, for this wise and wonderful young girl brings back a poignant delight in the infinite human spirit."

Levin's review ends with a hook designed to drag the reader out of his chair and straight to the bookstore when it opened on Monday morning. Which is exactly what happened. Who can say how many of us would have read Anne Frank if not for Meyer Levin's review, a brilliant, if suspect, piece of publicity that got the diary into the shops, into the hands of readers, and onto the desks of the Broadway producers who would engineer his downfall?

FROM the start Levin believed, and told Otto, that he was the ideal choice to turn the book into a play. He was Jewish, he felt a powerful sense of Jewish identity and of responsibility to the victims of the Nazis, he had been to the camps, he was among the diary's earliest and most devoted fans. But as the book's reputation and popularity grew, names more famous than Levin's began to be mentioned in connection with its theatrical adaptation.

More than fifty years later, one can hear the tone of the conversation change as the principals—Otto Frank, Meyer Levin, the publishers and producers—realized that a hot property was about to get hotter. Doubleday asked Otto if they could negotiate the theatrical rights for a 10 percent commission. Meyer Levin agreed. If Barbara Zimmerman was Otto's lost daughter, Levin was still the loyal, supportive nephew, and no one had forgotten the importance of his review. Confidently, Levin agreed to let the publisher make the best deal. Otto wanted everyone to be happy—for Levin to be rewarded and feel included, and for his publisher to maximize the book's potential without distorting Anne's work.

It was at this point that Levin wrote Otto Frank a fateful

letter reiterating his desire to do the adaptation. He had no interest in an agent's commission. All he wanted was to remain Otto's first choice as writer. Betraying a fatal failure of the imagination, Levin volunteered to step aside in the event that his withdrawal was the only way that a famous dramatist would agree to adapt the play.

Otto Frank's consent to let Doubleday sell the dramatic rights stipulated that the sale be approved by Levin. And even if Levin was beginning to seem like a troublemaker, everyone still imagined that everything would work out. Doubleday offered Levin half of their 10 percent agent's commission, and again Levin agreed.

Otto was reassured. The book would find the most prestigious and sensitive producer, Levin would either write, or collaborate on, the script. But one by one, the principal parties began to wish that Meyer Levin would just go away. In a letter to Frank Price, Barbara Zimmerman wrote that Levin's motives didn't seem mercenary, but that he was "screwing up the whole deal." In subsequent letters, Zimmerman, clearly at the end of her tether, described Levin as "impossible to deal with on any terms, officially, legally, morally, personally" and claimed "he seems bent on destroying both himself and Anne's play." Levin couldn't help but sense the growing disaffection with his role in the negotiations, and he slowly began to see himself as a desperate character, a failed writer, a poor Eastern European Jew surrounded by rich German-Jewish snobs, personified by Otto Frank.

In fact, Otto had been in a shaky financial position for much of his life, struggling to shore up a failing bank, and, with his cousin's help, running a modest dealership in jam-making products. Hitler had taken everything and murdered his wife and children. On his return to Amsterdam, he'd once more had to rely on the loyalty and charity of his former employees. And now it seemed that his daughter had not only given the world a literary classic but had provided him with a way to finish out his life in security and comfort. Who could blame him for wanting that after all he'd been through?

As the diary remained on the best-seller lists, the discussions about the selling of the dramatic rights began to include the magic words that sealed Meyer Levin's doom: *Famous writer. Important playwright.* Or sometimes, more coarsely, *big name.* The big names bandied about included Arthur Miller, Lillian Hellman, Thornton Wilder, Maxwell Anderson, Harold Clurman, Elia Kazan, and Joshua Logan. Three weeks after the book's publication, *Variety* ran a list of possible producers.

At first Levin's resistance was gentle. Every time a famous playwright was suggested, Levin helpfully explained why this or that celebrity was wrong for the job. Levin claimed that, as early as 1950, Otto Frank had agreed that Levin would serve as his American agent, negotiate the theatrical rights to the diary, and write the dramatic version. To which Frank replied that their understanding was not as formal and binding as Levin believed.

Otto reassured Levin of his faith in him and in Doubleday. But the first notes of desperation began to sound in Levin's letters to Otto. His advice grew more unreasonable and manipulative. He warned Otto that a famous playwright might leave too heavy a stamp on the material, and besides, some of the writers mentioned were has-beens with strings of failures. A few of the playwrights and producers Doubleday was considering had been investigated by the House Un-American Activities Committee; such candidates should be avoided lest the diary be caught in the political crossfire.

When Maxwell Anderson was suggested, and refused to collaborate with Levin, or with anyone else, Levin panicked. Anderson's "heavy-handed" work, Levin wrote Otto, had fallen out of favor. And his reputation might overshadow the diary: "If he writes the play, it will undoubtedly be known as Maxwell Anderson's play about the little girl, what was her name? It seems to me that the play should be identifiable as Anne's work." Finally, Levin claimed, Anderson was unqualified because he was not a Jew.

Otto Frank, who would spend the rest of his life ensuring that people not be judged and excluded on the basis of their color or race or religion, was offended by Levin's suggestion that non-Jews need not apply. In June 1952, he wrote back: "I always said that Anne's book is not a warbook. War is in the background. It is not a Jewish book either, though Jewish sphere, sentiment and surrounding is the background. I never wanted a Jew writing an introduction for it. It is (at least here) read and understood more by gentiles than in Jewish circles. I do not know, how that will be in the USA, it is the case in Europe. So do not make a Jewish play out of it! In some way of course it must be Jewish, even so that it works against anti-Semitism. I do not know if I can express what I mean and only hope that you won't misunderstand."

Otto was right to worry that he might have failed to explain himself clearly. His letter is filled with contradictions—Anne's story is Jewish, it isn't Jewish, it shouldn't be a Jewish play but should combat anti-Semitism.

Another recurring debate surrounding the diary has centered on the Jewish identity of Anne Frank. Did she think of herself as Jewish? Was she aware of what was happening to her fellow Jews? The answer is yes, and yes. Did the Franks keep a kosher house? A diary entry is devoted to Mr. Van Pels's expertise in making pork sausage. Did she believe in God? One of the things that surprised Otto Frank when he read his daughter's diary was how often, and how fervently, she wrote about God. Were the Franks "assimilated"? Yes. They celebrated Hanukkah *and* St. Nicholas Day. When Otto Frank asked Kleiman for a copy of the New Testament so that Anne could learn about it, and a "somewhat perturbed" Margot asked if he meant to give Anne a Bible for Hanukkah, Otto agreed that it might be better as a present for St. Nicholas Day, since—as Anne writes in her revisions and as Otto keeps in the published diary—Jesus didn't seem to go with Hanukkah. Like many modern parents, the Franks had, as their gods, their children.

Of course Anne's story was Jewish, and, despite his assimilated German-Jewish background, so was her father. His objection to making it "a Jewish play" was too complex to chalk up to his upbringing, his views on religion, or the postwar desire to return to normalcy and forget the catastrophic singling out of the Jews. It must have begun to occur to Otto that Anne's story could reach a much wider audience than the drastically reduced Jewish population. Of course the diary was read by more European gentiles than by Jews; there were so few Jews left. And even the forebearing Otto was clearly irritated by Levin's "nationalistic" objection to Maxwell Anderson. Beneath Otto's unease is an instinctive shrinking back from Levin's impulse to control him and the diary. As it rapidly became clear that Levin was not an impartial adviser but had a fierce personal stake in the diary's future, Doubleday began to leave him out of the loop.

Meanwhile, Levin was busily writing his own dramatization of the diary, which was finished by the time Doubleday engaged a producer, a seasoned Broadway veteran named Cheryl Crawford. Initially, Levin liked Cheryl Crawford, though again there seems to have been a misunderstanding about his role. Crawford initially suggested that Lillian Hellman and Clifford Odets adapt the diary, but when she heard about Otto's loyalty to Levin, she gave Levin two months to come up with a draft.

That summer, Levin worked on the play, taking a break to write a half-hour radio adaptation of the diary for the American Jewish Committee, which was slated to be broadcast in September. During this time, another producer—Kermit Bloomgarden—suggested himself for the project and proposed that Arthur Miller write the adaptation. At least Miller was Jewish, said Levin. But he wasn't Jewish *enough*.

In the fall of 1952, Otto arrived from Europe—on Yom Kippur, as it happened. Barbara Zimmerman met his boat at the pier. The next day, Cheryl Crawford told Levin that she liked his first draft. But two days later she informed him that she had reread it in the middle of the night and now had serious reservations. She agreed that he should be given time to revise it. Too little time, he argued. That Levin was given a weekend to rework the entire thing suggests that Crawford had already decided against him. The playwrights who were chosen would eventually write eight drafts.

With Doubleday's encouragement, Otto Frank engaged a lawyer, Myer Mermin. Though Mermin determined that Levin had no formal rights to the play, he did acknowledge that his casual arrangements with Otto might support a legal challenge. Mermin suggested that Levin be given a month to submit his script to an approved list of producers. If none was interested, Levin would renounce all rights to the drama.

When no willing producer could be found, Levin argued that his play was being discriminated against for being too Jewish, a quality that offended the "doctrinaire" political sensibilities of the anti-Semitic Stalinists, among them Lillian Hellman, who was advising the production team. The period in which this censorship occurred, Levin would later write, coincided with the height of Stalin's campaign against Jewish writers and cultural figures. Not wanting to give the House Un-American Activities Committee more fuel for its fires, Levin

claimed, he kept silent about the reason for his work being suppressed and wound up pinned between Josef Stalin and Joseph
McCarthy. Others, including Crawford and the play's eventual
producer, Kermit Bloomgarden, blamed Levin's play for being
insufficiently dramatic to work on the stage. Levin argued his
case in letters that the *New York Times* and *Variety* refused to
print, and in an increasingly testy correspondence with Otto
Frank. In one letter to Otto, Levin called Cheryl Crawford a
"castrating homosexual."

By now a larger cast of characters had been assembled to
weigh in: Hellman, Bloomgarden, Elia Kazan. Here is where
Levin (and Ralph Melnick) see the heavy fist of Lillian Hellman crashing down as she arm-wrestles the production away
from Levin, with his narrowly sectarian (that is, Jewish) slant
on Anne's story, and conspires to hand it over to a team who
will be more in harmony with her own anti-Semitic, extreme-
left-wing program. Melnick finds abundant evidence supporting Levin's claims that Hellman manipulated the situation to
achieve her Stalinist agenda. But that seems unlikely, since, in
Lillian Hellman, personal ambition appears to have trumped
politics. Lillian Hellman wasn't Stalin's agent, but her own. She
knew the most important writers, she knew theater people, she
could make things happen in a way that Meyer Levin could
not.

The other so-called co-conspirators and dupes—producers,
writers, directors—intuited, early on, that the play might turn
a profit. Even if the principal architect of the "cabal" against
Levin was a Communist, rarely has Communism achieved such
capitalist cash-cow success, and it seems unlikely that Hellman
and her cohorts conspired to funnel the profits from Broadway
to the Kremlin. Regardless of the political views of the players,
the drama of the diary's adaptation was not about Communism
but about capitalism working exactly the way it's supposed to,

cutting the dark stuff, the Jewish stuff, the depressing stuff, emphasizing the feel good—and making money. This was 1950s America, the war was over, the "healing" well under way, and it was time for the sitcom teen, together with Mom and Dad and Sis, to head off to the secret annex.

In early conversations between Lillian Hellman and Garson Kanin, who would eventually direct the Broadway production, Hellman made a telling remark that would probably have been lost on Levin had he heard it. While acknowledging the diary's literary importance, Hellman explained that she was the wrong person to adapt it. Such an adaptation, she said, would be so depressing that the producers would be lucky if the play ran for one night. They needed a playwright with a "lighter touch." Later, the chosen writers—Frances Goodrich and Albert Hackett—were instructed to emphasize Anne's humor. "The only way this play will go will be if it's funny," advised Kermit Bloomgarden. "Get (the audience) laughing . . . That way, it's possible for them to sit through the show."

Once again, Levin got it wrong. His approach to Anne's diary may have been "too Jewish" but it was also, more problematically, too serious. Ironically, the ponderousness and sententiousness of his own adaptation meant it might have had a better chance of being produced in the state Socialist system he despised and feared, a climate in which art, however wedded to propaganda, was divorced from commerce, and less dependent on a theatergoing public who could choose to buy, or not buy, expensive Broadway seats. Only in a capitalist society were ticket sales related to a play's ability to make its audience feel chastened but uplifted, sad but hopeful. The play's producer and director understood that a drama confronting the horrors of the Holocaust and the political ramifications of Nazism, Zionism, and Jewish spirituality was unlikely to pack the house night after lucrative night. Whatever the strengths and virtues of the play that Meyer Levin wrote, and that can still be read in a version performed by the Israel Soldiers Theatre, levity and humor were not among them.

AT SEVERAL points in *The Obsession*, Levin the novelist takes over from Levin the memoirist and Levin the polemicist, and we read scenes, drawn from life, that suggest a different explanation for his disappointments than the one that he is (or believes he is) providing. Several excruciating passages detail his bruising encounters with Joseph Marks, the vice-president of Doubleday, who was deputized to handle dramatic rights to the diary.

In his elegant office, Marks reads a list of famous producers who have made offers and informs Levin that the deciding factors will be the track record of the producer and the fame of the adapter. Obviously, he is saying this to a virtually unknown writer. Levin points out that the diary is his project, and Marks says he understands that, but a "big-name dramatist would virtually assure a Broadway success." Again, it's difficult to link this conversation to the Stalinist ideology that, Levin claims, engineered his failure. On the other hand it's all too easy to imagine the unease of potential backers listening to Levin's plans for the adaptation: "The very origin of our theater was in religious plays of martyrdom . . . the play, if done, must be a reincarnation. In the persistence of the living spirit each spectator would feel a catharsis. When the spirit reappeared before him, indestructible, the crematorium was negated . . . I saw the form almost as a ballet, a young girl's probing, thwarted at each impulsive moment while she strives for self-realization."

To read Meyer Levin's adaptation of the diary is to confront the pitfalls of basing a work of art on Big Ideas: martyrdom, reincarnation, self-realization. As Levin's drama begins, a group of mourners in black raincoats chant the Hebrew prayer for the dead. A narrator, employed throughout to apprise the audience of historical developments—the implementation of the final solution, the construction of the camps at Auschwitz and Treblinka, the German incursions into Russia and Africa, the Allied invasions, as well as the shifts in time between the acts—outlines Otto Frank's background and his emigration from Frankfurt to Holland.

The prologue opens on an Amsterdam street, outside the Franks' home. It's Anne's thirteenth birthday, and she and her friends, Joop and Lies, are discussing her gifts. The conversation shifts to Anne's flirtation with a certain Harry Goldberg, and Joop asks if Harry is taking Anne to the Zionist meeting. The girls discuss the restrictions that the Nazis have imposed on the Jews and the dangers of violating curfew. One says, "Quite a triumph, to get yourself sent to the concentration camp in Westerbork for ten minutes extra in the company of Harry Goldberg."

Otto Frank appears, carrying a parcel that he claims to be giving to Dutch friends for safekeeping from the Nazis. He and Anne discuss the possibility of going into hiding. There's a long conversation about the onset of menstruation, and Mrs. Frank laughs at Anne for hoping that her first period might arrive as a birthday present. Then the call-up notice comes, spoiling the amiable mood, and Mrs. Frank says, "We must go into hiding at once."

Cut to the secret annex, where Margot is telling Miep about the guilt she feels for endangering her family by ignoring the summons; she mentions her dream of someday moving to Palestine. The rest of the family arrives, there are intimations of Anne's conflict with her mother, followed by a romantic scene between Anne and her father in which Anne speaks of her desire to be alone with him in the world.

The entrance of the Van Daans is more artfully handled than in the Goodrich-Hackett version, in which we are meant to accept the preposterous notion that the two families are meeting for the first time. In Levin's play, as in life, they have a history. There's a nice moment when Mr. Koophuis asks Mrs. Van Daan to remove her high heels, so as not to make any noise, and she says, "Never let a man choose a house."

Likewise, when Dussel appears in the second scene, his characterization is closer to what we know of Fritz Pfeffer than is the buffoon who stands in for him in Goodrich and Hackett's drama. As the others wait for Dussel to show up, Mrs. Van Daan says that she hears he's the biggest Don Juan of all the dentists in town; perhaps his delayed arrival means his female patients are reluctant to let him go. There's an argument in which Anne objects to letting Dussel share her room, and another about which books children should be allowed to read—a disagreement that appears in the diary.

The extended debates about spirituality and Palestine ("It's part of being something more than yourself. . . . It's making a free life for ourselves," says Margot) point out the strength and weakness of Levin's play. Anne is given at least some of the intelligence she displays in the diary, and which the Goodrich-Hackett heroine lacks. In one scene with Peter, she quotes Plato about men and women having once been united, then splitting into two halves, each struggling to find completion. One can hardly envision the Anne who pouted and pranced her way across the Broadway stage citing *The Symposium.* Yet the long passages of dialogue with which Levin delineates Anne's character contribute to the static, discursive—*unstageworthy*—quality that troubled the producers. One can imagine potential backers cringing when Anne and Peter talk about how to determine the gender of a cat.

Throughout, Levin's play makes one realize how little "action" is in the diary, and how much is domestic, interior, and psychological. Having failed to discover the rhythm of crisis, danger, and (relative) relaxation that would engage the audiences who attended the Broadway production—a tension that Goodrich and Hackett labored, in draft after draft, to sustain—Levin relies on frequent mentions of Westerbork to create a sense of threat. The break-in downstairs is more discussed than dramatized, though there is one tense moment when a bomb falls nearby.

In a scene between the Franks, Edith—whose piety becomes so oppressive that you can understand what irritated Anne, though this seems not to have been Levin's intention—wonders if God is punishing them for not having taught the children more about their religious heritage. When Anne announces her plan to write a book entitled *The Hiding Place*, her sister replies, "I suppose it could be exciting. All the times we've been nearly caught, the robbery in the front office, and that nosey plumber"—events that Levin fails to exploit for their drama, though later, during the Hanukkah party, we see two thieves picking the downstairs lock.

For Levin, a long scene in his play—in which Anne explains her religious feelings to Peter—conveys the soul of Anne's book. In the diary, Anne's first doubts about Peter, misgivings that upset her even in the heady phase of their romance, occur when he remarks that life would have been easier had he been born Christian. In later entries, Anne's reservations about Peter often surface following spiritual discussions; she is troubled by his distaste for religion in general, and for Judaism in particular.

Hans, the object of Cady's romantic feelings in Anne's novel-in-progress, *Cady's Life*, is more religious than she is, and offers her this guidance (in a scene that goes on even longer than the conversation in Levin's play) when they meet near the sanitarium where she is recuperating:

"When you were at home, leading your carefree life . . . you just hadn't given God a lot of thought. Now that you're turning to Him because you're frightened and hurt, now that you're really trying to be the person you think you ought to be, surely God won't let you down. Have faith in Him, Cady. He has helped so many others."

One can imagine Levin being less than thrilled by the pantheistic (or animistic) beliefs that Hans expresses: "If you're asking what God is, my answer would be: Take a look around you, at the flowers, the trees, the animals, the people, and then you'll know what God is. Those wondrous things that live and die and reproduce themselves, all that we refer to as nature— that's God." In fact, this conflation of nature with God runs close to the core of what Anne appears to have believed during her final months in the annex.

In any case, if Anne and Peter's extended metaphysical discourse seemed less gripping to others than it did to Levin, that may have had less to do with his play's "excessive Jewishness" than with the producers' pragmatic realization that, if an audience is going to watch two teenagers on stage, it's not because they're having a conversation about God.

Like the play that ran on Broadway, Levin's drama ends with the quote about people being good at heart, but, unlike the Goodrich-Hackett version, it puts the line back in context, amid the dialectic between Anne's hope and her terror that the world will turn into a wilderness. The stage goes dark, the sounds of combat come up.

"End of the diary," declares the narrator. The stage directions specifiy that the battle noises grow louder, but reading the play, one may be more likely to hear the responses of the producers to whom Levin showed his drama: Dark. Depressing. Jewish. Gloomy. Insufficiently *universal*. The polar opposite of *commercial*.

EVEN if Cheryl Crawford had admired Levin's script, it would have been hard for her to ignore the siren song playing in everyone's ears. *Lillian Hellman believed* they could get a playwright who was not only more famous but classier than Meyer Levin.

Among the classiest names being mentioned was that of Carson McCullers, who had adapted her novel, *The Member of the Wedding*, for the stage, to great success and acclaim. The fact that its heroine was a teenage girl was one of the reasons, Barbara Zimmerman told Otto Frank, why McCullers was the perfect choice. Frank Price, at whose Doubleday Paris office the diary had been rescued from the rejection pile, contacted McCullers, then living in France with her husband, Reeve.

Carson McCullers wrote Otto Frank, "I think I have never felt such love and wonder and grief. There is no consolation to know that a Mozart, a Keats, a Chekov is murdered in their years of childhood. But, dear, dear Mr. Frank, Anne, who has had that . . . gift of genius and humanity has, through her roots of unspeakable misery, given the world an enduring and incomparable flower. Mr. Frank, I know there is no consolation, but I want you to know that I grieve with you—as millions of others now and in the future grieve. Over and over in these days I have played a gramophone record of the posthumous sonnets of Schubert. To me it has become Anne's music . . . I can't write an eloquent letter but my heart and my husband's heart is filled with love." She added that she hadn't given much thought to the idea of writing a play based on the book. "I have only read the diary and am too overwhelmed to go any further."

Otto and his second wife, Fritzi, visited Carson and Reeve. Soon after, McCullers wrote Fritzi, "We have no formal religion but there are times when one understands a sense of radiance—and that feeling was with us when Otto was in our home. What is this radiance, this love? I don't know, I only want to offer our joy that you and Otto are united and this carries all our love to you."

Two months later, Carson McCullers decided not to proceed with the project. "In spite of our deep feeling about Anne's diary, it requires more technique in the theater than I can command . . . You see it is different doing a solitary work—that is all I have done—than adaptation on others' books. Consequently I feel the result might lead to unhappiness to all concerned." Later, she would claim she feared that immersing herself in the diary might damage her already fragile health, and that the mere prospect of it had caused her to break out in hives.

In April 1953, Cheryl Crawford, alarmed by Levin's increasingly litigious threats and demoralized by the financial loss she'd sustained staging Tennessee Williams's *Camino Real,* withdrew from the negotiations. That fall, Kermit Bloomgarden signed on to produce. Though Bloomgarden showed little interest in Levin's adaptation, Levin behaved as if Bloomgarden's involvement signaled a new beginning. When Levin realized that Bloomgarden was not the ally he had hoped, his behavior further deteriorated. In a letter to Otto, Levin claimed that his passion for his play was exactly like Otto's feelings for his daughter, and expressed his conviction that neither the play nor the child should have been killed "by the Nazis or their equivalent."

It was around this time that Bloomgarden first contacted the husband-and-wife screenwriting team Frances Goodrich and Albert Hackett. They'd had glamorous careers in Hollywood, where their hits had included *Seven Brides for Seven Brothers, The Virginian*, and *Easter Parade*. "The Hacketts of Hollywood," Levin called them. They'd come with the highest credentials, recommended by Lillian Hellman who, as we have seen, felt that the diary needed just the sort of light touch that the Hacketts could provide. As the authors of *Father of the Bride*, they had demonstrated their ability to write about adolescents, just as the experience they'd had in doctoring the script of *It's a Wonderful Life* had proved that they were able to brighten "dark material." Who could balance charm and suspense? The adapters of Dashiell Hammett's *The Thin Man*.

Goodrich and Hackett hesitated, but were at last persuaded by the possibility of leavening the tragic story with "moments of lovely comedy which heighten the desperate, tragic situation of the people." They saw it as a "tremendous responsibility" and were flattered by the invitation to be associated with a book that a major figure like Lillian Hellman considered serious literature. They even agreed to take a pay cut and to accept a fee far below that which they were accustomed to receiving for screenplays. In fact, they were working "on spec," an almost unheard-of situation for professionals of their stature. They would get a thousand dollars if, and only if, Bloomgarden picked up the script.

They wrote Otto Frank to say that they felt honored to have been chosen to bring his daughter's spirit and courage to the stage, and Otto wrote back, pleased that they had been so moved by the diary and offering his help. The Hacketts were less successful with their conciliatory note to Meyer Levin, which elicited a four-page, single-spaced disquisition on how badly he had been treated. When the Hacketts began their research, visiting Jewish bookstores and a rabbi in Los Angeles, the frosty receptions they received made them worry that Meyer Levin had managed to turn the community against them.

On January 13, Meyer Levin placed the following paid advertisement in the *New York Post*:

A Challenge to Kermit Bloomgarden

Is it right for you to kill a play that others find deeply moving, and are eager to produce?

When you secured the stage rights to Anne Frank's Diary of a Young Girl you knew I had already dramatized the book, but you appointed new adapters . . . and shoved my play aside. The Diary is dear to many hearts, yours, mine, and the public's. There is a responsibility to see that what may be the right adaptation is not cast away.

I challenge you to hold a test reading of my play before an audience.

A plea to my readers.

If you ever read anything of mine . . . if you have faith in me as a writer, I ask your help. Write to Mr. Frank and request this test.

My work has been with the Jewish story. I tried to dramatize the diary as Anne would have, in her own words. The test I ask cannot hurt eventual production from her book. To refuse shows only a fear my play may prove right. To kill it in such a case would be unjust to the Diary itself.

The question is basic: who shall judge? I feel my work has earned the right to be judged by you, the public.

Write or send this ad to Otto Frank . . . as a vote for a fair hearing before my play is killed.

Levin's plea had the unintended effect of finally alienating Otto. Bloomgarden wrote the Hacketts, telling them that he would refuse to dignify Levin's challenge with a reply. As further evidence of Levin's disreputable character, Bloomgarden cited the fact that Levin had reviewed, in the *New York Times*, a book that he was representing, as its agent.

The Hacketts began work on the play. Writing eight drafts would involve great strain for both writers, elicit copious tears from Goodrich (weeping she ascribed to guilt over not having known and done more about what happened to Anne and others like her), and spark numerous marital squabbles, some private, some public. In addition, they wondered whether, at a time when the United States was interested in cultivating Germany as an ally in the Cold War and as a market for American investment abroad, anyone would want to stage a drama that accused the Germans.

They scrapped the second draft when they realized that their fear of making the characters unsympathetic had kept them from making them human. Encouraged by having found an apparently workable ending, they sent their fourth draft to Bloomgarden and Hellman, both of whom hated it. Their spirits were further dampened by a letter from Otto Frank, who said that he could not approve a play that ignored Anne's idealism, her moral vision, and her desire to help mankind. Oddly, what Otto seems to have wanted was something more like Meyer Levin's rejected version. Otto complained about the "snappish" characterization of Margot, criticized the downplaying of Anne's friction with her mother, and doubted that their adaptation would appeal to young people.

Meanwhile, royalties from the sales of the book had allowed Otto and Fritzi to move to Switzerland, where some of Otto's family lived, having taken refuge there before the war. Despite his reservations about the Hacketts' early efforts, he was relieved that the project was going forward. He reconciled himself to the fact that some of the changes he proposed—during the Hanukkah scene, the men should wear hats—were approved, while others—the actors should sing the solemn, traditional hymn in Hebrew rather than the raucous party song in English—were ignored. Bloomgarden also chose not to follow Otto's suggestion that the theater program contain a note stating that the play was based on actual events. Later, Otto would hear from a Dutch acquaintance that at one performance she had sat beside an American woman who had seen the play three times without having any idea that the actors were portraying real people.

Otto's fragile peace was regularly broken by progressively more disturbing communications from Meyer Levin. For a while, Otto continued to defend himself on the subject of the play's Jewish content, but later, probably on the advice of his lawyer, he circulated a statement expressing his confidence in his own ability to interpret and stand up for his daughter's ideals and example.

Levin was unconvinced. He wrote to Otto questioning his right to decide how Anne would have wanted to be portrayed, accusing him of foisting his own interpretation on an unsuspecting public, and invoking Anne's hatred of injustice to suggest that she might have sided with Levin, the victim of injustice, against her father, the perpetrator. Though he admitted that Otto might have known her as a daughter, Levin insisted that Otto could not have known her as Levin did, the way one writer knows another. That deeper intimacy, Levin claimed, should give him the right to decide—in Anne's name—who should adapt her diary. And he was the person to do it.

Meanwhile, Otto continued to send amiable letters to Levin's wife, who seems to have hoped, as did Otto, that the two friends could be reconciled. Levin responded with the most direct attack so far, claiming that Otto's treatment of him was typical of the "cavalier" way in which Otto used and discarded his allies; Levin cited the example of one of the diary's early translators, whose work Otto had decided against.

Even as we recoil from Levin's claim to speak for Anne, something keeps compelling us to see things from his side, or at least to understand what made it so hard for him to give up. In his writings, he repeatedly emphasizes that he was present when the camps were liberated, and that the memory of the dead prevents him from standing by and witnessing Anne's transformation into just another teenage girl. But of course Otto Frank also had direct experience of the camps, not as a liberator, but as a prisoner—a fact that Meyer Levin appeared to forget as his obsession spun further out of control.

In September 1954, as they struggled with the fifth draft of the play, the Hacketts consulted Lillian Hellman, who made a number of structural suggestions that Frances Goodrich called "brilliant." Garson Kanin, whose triumphs included the popular Katharine Hepburn–Spencer Tracy films *Woman of the Year* and *Adam's Rib*, and the Broadway hit *Born Yesterday*, was hired to direct. It was Kanin's idea to end the play with Anne's statement about people being good at heart and to ramp up the tension by adding threatening noises—footsteps, sirens—from outside the attic.

Kanin advised Goodrich and Hackett to eliminate a conversation in which Peter expresses his outrage at the fact that they are suffering because they are Jews, to which Anne replies that, throughout history, Jews have always had to suffer. Kanin reminded the playwrights that every minority has experienced its share of persecution, and that for Anne to single out the Jews "reduces her magnificent stature." Without such "embarrassing . . . special pleading . . . the play has an opportunity to spread its wings into the infinite."

The Hacketts continued to produce drafts that disappointed Bloomgarden, who advised them that the romance with Peter was not intense enough, that the Anne they'd created was not enough like George Bernard Shaw's Joan of Arc, that the character of Mrs. Van Daan was insufficiently shrewish, that Anne's relationship with her father needed to be more loving, and that they were turning Anne into a sour and pessimistic young woman. First the Hacketts were told that their play was too dark, and then that it was not dark enough.

In the fall of 1954, they met Garson Kanin in London, and the three of them spent long days collaborating on yet another draft. In December, they visited Otto Frank in Amsterdam for a week that Frances Goodrich described as "very harrowing."

"I thought I could not cry more than I had," she wrote. "But I have had a week of tears." A photo shows the playwrights, Kanin, Otto Frank, Johannes Kleiman, and Elfriede Frank standing in front of 263 Prinsengracht. Kleiman and Fritzi wait patiently in the background, while their American visitors look suitably chastened and ennobled by the chance to walk in the footsteps of the girl who had so inspired them.

Dispatched to do research at the secret annex, a photographer documented every inch of the attic. Recordings were made of ambient street noise and the Westertoren bells. Goodrich described stretching out her arms in the room that Anne had shared with the dentist, while Kanin noticed that one of the pictures Anne had on her wall was a still photo of Ginger Rogers in *Tom, Dick and Harry*, a film he'd directed. Clearly, the project was meant to be.

Though the pictures on the walls of Anne's room are occasionally rotated by the museum staff, the postcards, snapshots, and newspaper clips that decorate the room today are more or less the same ones that the American theater people must have seen. Yet the finished play suggests that the Hacketts and Garson Kanin factored only a few of those images into their version of Anne. They captured the starstruck Ginger Rogers fan, the giddy teen with a fondness for the royal princesses and Deanna Durbin, but seem to have missed the ironic humor of the child amused by the chimpanzee tea party, as well as the adolescent eroticism of the girl drawn to the languid Jesus in Michelangelo's *Pietà*.

Their understanding of Otto was equally skewed and incomplete. "In all my meetings with him," Kanin said of Otto Frank, "he was unhurried, casual, old-worldish. He talked about the hide-out and the arrest without an ounce of emotion. 'This is a cold fish,' I told the Hacketts." But Kanin changed his opinion on learning that Otto had collapsed after the American theater people left Amsterdam. "He had been crushed, but he had not shown it. He had been as he had been in the days when the Gestapo was outside the door—a tiny, tiny, modern miniature Moses. If he had shown a moment's fear then, the whole annex would have crashed down."

A few weeks after Kanin's trip to Amsterdam, Meyer Levin, who had found a lawyer willing to take his case, filed suit in New York State Supreme Court against Cheryl Crawford and Otto Frank, charging them with breach of contract. He sought a monetary award of $72,500 from Crawford, while from Frank he asked that they forget the damage each had inflicted on the other and return to the point at which it had been understood that Levin would write the adaptation. Otto sent the Hacketts a reassuring letter. How this would have upset Anne, Otto wrote—Anne, who, like him, hated quarrels.

Otto Frank's lawyer managed to have the suit set aside (but not dismissed) on a technicality: the summons could not be delivered to Otto, who was in Switzerland. Levin suffered an emotional collapse, but nonetheless found the strength to send Otto a letter vowing to fight the production, a struggle he compared to the Warsaw ghetto uprising.

MEANWHILE, in New York, the play was being cast. Joseph Schildkraut was chosen to play Otto, despite some hesitation on the part of Bloomgarden, who—according to Schildkraut—could not dispel his impression of the actor as the "flamboyant and dashing" character he had played in previous roles. In her

New York Times piece about the production, Frances Goodrich reports having noted, on first meeting Otto Frank, his "uncanny" resemblance to Schildkraut. Susan Strasberg was picked for Anne, while Mrs. Frank would be played by Gusti Huber, an Austrian actress alleged to have acted in Nazi propaganda films. (*Echoes from the Past,* a documentary about the making of the 1959 Hollywood version of *The Diary of Anne Frank* that appears as supplemental material on the DVD, explains that Schildkraut and Huber, who re-created their stage roles on screen, were both Austrian and both "had first-hand experience of Nazi anti-Semitism.")

Alarmed by the rumors about Huber, the Hacketts wrote Otto Frank, who replied that his Viennese wife had not heard of Huber but was curious about her. Otto seems not to have followed through on his offer to inquire further into Huber's background.

Dennis Hopper was the Hacketts' first choice for the role of Peter, but Warner Brothers, with whom Hopper was under contract, insisted that he remain on the set of *Giant,* where he was being considered as a replacement for the temperamental James Dean. Lou Jacobi was brought in to play Mr. Van Daan, and Jack Gilford was cast as the dentist, Dussel.

Rehearsals began in late August, and at the first run-through, Kanin gave his actors an inspirational talk. "This is not a play in which you are going to make individual hits. You are real people, living a thing that really happened."

This hortatory speech had only a limited effect on his cast. Strasberg was, it was felt, an ingenue-diva, childish, spoiled, and reluctant to take direction, while Schildkraut had problems tamping his ego down enough to play the gentle, mild-mannered Otto. Initially, the actor resisted the suggestion that he shave his leonine hair so that he would more closely resemble the balding Mr. Frank, but subsequently discovered that this blow to his vanity provided the key that gave him access to Otto's character. Kanin used his insight into Otto Frank as a "tiny, tiny miniature modern Moses" in the direction he gave Schildkraut. "I told him also, you, too, can collapse the way Mr. Frank did, but only after the curtain comes down. We worked out a Mr. Frank who does not show how he feels. But we hope the audience will sense his strength."

The playwrights and the director were troubled by problems with the second act and worried by unpromising advance ticket sales. "Both Kermit and Gar talked their heads off," Goodrich confided in her diary. "No good. Too serious."

They decided to increase the suspense in the second act by inventing a fictional scene in which Mr. Van Daan is caught stealing bread. This plot turn aroused Otto Frank's fervent objections, to which no one paid much attention. What disturbed him was not so much its lack of veracity or plausibility, but rather the fact that his former business associate, friend, and roommate had living relatives whose feelings might be hurt.

A few days before *The Diary of Anne Frank* opened on Broadway, *New York Times* reporter Bernard Kalb helped theatergoers understand what they were about to see. The play, he wrote, is "partly" an account of eight Jews in hiding. "Mostly, though, it is the story of one of them—a young girl who refused to be robbed of the adventure of adolescence." Most of the article draws on an interview with Garson Kanin, in which Kanin describes the trip to see the secret annex and to meet Otto Frank in Amsterdam, and explains why Anne exits smiling. "Well, it seems that Anne's first reaction on finally leaving the annex was one of joy. At last, she was in the sunshine." Moreover, he says, when Otto Frank last saw his daughter, on the train on which she was being sent to Bergen-Belsen, she was smiling and waving at the crowd of men, hoping her father might see her.

" 'That's the way I last saw her,' Kanin claims Otto remembered. 'Smiling and waving. She never knew that I saw her.' " Kanin continues, "I told this, the last eye-witness account of Anne, to Susan, and that's why Anne is smiling when we last see her. I couldn't have thought that up. It would never have occurred to me. Yet that is the essence of Anne."

As we know, this was hardly the last sighting of Anne. A production that ended with the reports of the women who saw the emaciated, dying girl at Bergen-Belsen would have been a different play from the one Kanin was directing, which was "not a war play, or even a sad play." As Kanin told the *Times* reporter, "This play makes use of elements having mainly to do with human courage, faith, brotherhood, love and self-sacrifice. We discovered as we went deeper and deeper that it was a play about what Shaw called 'the life force.' Anne Frank was certainly killed, but she was never defeated."

The play opened at the Cort Theater on October 5, 1955. Otto Frank declined to attend the premiere because he feared that it would be too painful to see himself, his wife, and his children portrayed on stage. In addition he had been warned by his lawyers to stay out of New York to avoid being served with a legal summons. As a consequence of Meyer Levin's lawsuit, all of Otto Frank's royalties had been put into escrow until the case was settled.

The drama was not only a critical, but also a popular, success. It ran for 717 performances over nearly two years, and went on to win the Pulitzer Prize and the New York Drama Critics Circle Award.

Meyer Levin stood outside the circle cast by the brilliant spotlight that could have shone on him. He continued to write abusive letters to Otto Frank, equating the fate of his play with that of the murdered Jews. In December 1956, he again filed suit in the supreme court of New York, this time seeking $200,000 in damages and adding charges of plagiarism to his previous claims of breach of contract. Once again, the contradictions and confusions surrounding the diary in general—and Levin's case in particular—surfaced as Levin, who had been arguing that the play was a travesty of the diary, now insisted that it was based too closely on his adaptation.

Even as the judge dismissed the charges of fraud and breach of contract, he allowed a jury to determine the validity of the plagiarism charges. They ruled that Levin should receive $50,000 from Kermit Bloomgarden and Otto Frank. The verdict was set aside by the court, sensibly ruling that plagiarism was difficult to determine when the works in question were based on a common source. Eventually, after Eleanor Roosevelt had offered and then withdrawn her support for Levin, a settlement was reached, and Otto agreed to pay Meyer Levin $15,000. But still Levin kept writing to Otto, charging him with having returned evil for good and having betrayed his daughter. Which is the image that we're left with at the end of this painful story: a man possessed and maddened enough to write such letters, and a bereaved father receiving them, until at last he reached the point at which he refused to read any more.

LIKE Meyer Levin's adaptation, the Goodrich-Hackett play begins with a prologue, though here it serves as half of a framing device that book-ends the central action. The war is over. Broken, widowed, and childless, Otto Frank returns to the attic and, as the Dutch helpers look on, he begins to read aloud from the diary that Miep has just given him. In concert with Otto's, we hear Anne's voice, and then her voice takes over, describing the prohibitions, the special schools, the yellow stars, the events that occurred when "things got very bad for the Jews."

Shouldn't this have reassured Levin that Jewish religious and historical context had not entirely disappeared? In fact, the Hacketts went to such pains to explain the reason for the Franks' incarceration that what is integrated and organic in the diary seems awkwardly expository in the drama. But how could the Hacketts have told the story *without* Judaism and the final solution? What else were the Franks, the Van Daans, and Dussel doing in the secret annex? Regardless of what song they sing at the Hanukkah party, regardless of what language they sing it in, regardless of whether Anne laments the persecution of her people or of all people, we are never unaware that the characters onstage are Jews.

When Dussel arrives, we hear (as we do in the diary) that Jews are being rounded up and deported. But a note of unreality creeps into the drama whenever there is a mention of life beyond the attic walls. In the diary, the residents know perfectly well that they are condemned to stay in the attic for the duration of the war. But in the play there's some talk of the dentist remaining only until he can find somewhere else to go, just as later, after the scene in which Mr. Van Daan is caught stealing bread, Mrs. Frank suggests that the Van Daans find another hiding place. All of which makes it sound as if they are facing an extreme sort of housing crisis rather than trying to save their lives by evading the Gestapo.

If the scene in the Goodrich-Hackett play that most troubled Meyer Levin was the one in which Anne's observation about the sufferings of the Jews was generalized to include the sufferings of the human race, I'd nominate another episode as the drama's most distressing moment. It occurs early on, and involves our introduction to Anne. Poor Anne, so conscious of her self-presentation as she rewrote her diary to reflect the way in which she wished to be perceived! How embarrassed she would have been to learn that practically the first thing we see her do is remove her underpants, in full view of her fellow actors and the audience. When her mother objects, she replies that she has several more pairs underneath.

Like so many others, Meyer Levin seems not to have noticed that, if Anne's spirituality has been omitted, so has nearly everything else about her. Judaism is only one feature that was altered in the makeover that left the character of Anne Frank virtually unrecognizable as the author of the diary. On the page, she is brilliant; on the stage she's a nitwit. In the book, she is the most gifted and sharp-sighted person in the annex; in the play, she's the naive baby whom the others indulge and protect. For all her talk about being treated like a child and not knowing who she was, Anne saw herself as an adult and the others as children. In the drama, those relations have been reversed. Anne is always needing the obvious explained; she's invariably the slowest to grasp the dangers and necessities of their new life. A preteen trickster, she can't stop playing pranks, hiding Peter's shoes and saying lines like, "You are the most intolerable, insufferable boy I've ever met!" How the real Anne Frank would have cringed at the scene of her spilling milk on Mrs. Van Daan's precious fur coat, and how that brave girl would have railed against being shown fainting from terror when thieves break in downstairs.

Most critiques of the play, Levin's included, seem narrowly focused and myopic, oblivious to the ways in which Anne had been turned into a silly and shallow version of herself. Seriousness and humor were equally important to Anne, who by all accounts was a funny girl. But one can't quite imagine her being arch and kittenish, as she so often is in the drama.

For all the attention given to the question of what hymn would be sung at the Hanukkah party, scant mention has been made of the fact that Anne attends the celebration with a lampshade on her head. People who knew Anne describe her as a chatterbox and a showoff, and in her diary, she portrays herself that way. But no one has ever suggested that she was stupid, which is the impression created by scene after scene. It's hard to picture the real Anne exclaiming the Dutch or German equivalent of "Whee!", which, in the drama, is the sound she makes at the end of the workday when the annex residents are released from having to tiptoe around in their socks. In the script, many of Anne's lines end with an exclamation point.

The process of coming to take one's self seriously as a writer may be even less dramatic than that of embracing one's identity as a Jew, and yet one can't help wishing that the greatness of the diary itself—and the value that Anne placed on her work—had somehow been evoked to counterbalance her youth and innocence. In life, she received the diary several weeks before her family vanished from their old life. But in the play, she's given the journal when they are already in the annex. It's not the tweaking of history that grates so much as Anne's (and her family's) response. Delighted by the gift, she's all ready to run downstairs to get a pencil so she can start writing when her mother forbids her to go to the office. Only then, the stage directions tell us, does Anne understand what it means to have gone into hiding, when in fact we can assume that Anne grasped the implications when her father first mentioned the possibility.

Anne's response to a conversation about burning the diary to protect their helpers ("If my diary goes, I go with it!") is given, in the play, to Peter to say about his cat. (In the film, the line is restored to Anne to say about her diary, so that both Anne and Peter say, "If it goes, I go with it," a repetition that equates a literary masterpiece with a noisy pet.) When Anne tells Kitty that she wants to go on living after her death and wonders if she will ever be able to write well enough, the honest answer—were one to judge solely from the evidence offered by the play—would be a dubious *maybe*.

What could the girl we see in the play manage to write? When she reads aloud from her journal, stalls and ellipses interrupt every cogent reflection or opinion. The prodigiously articulate author can hardly utter a sentence without pausing to collect her scattered thoughts, none of them especially incisive. The pointed accuracy of her observations has been blunted, the delicacy of her perceptions has everywhere been coarsened. The beautiful line about Margot lacking the nonchalance for conducting deep discussions reappears in the play as a banal complaint about her older sister taking everything too seriously.

Anne's intriguing contradictions have been simplified out of existence. The diary's final entry, in which she writes about the gap between her inner and outer selves and speculates about what she could become . . . "if there weren't any other people living in the world" has been edited to remove the part about other people. Now her existential fantasy trails off. ". . . Some day, when we're outside again, I'm going to . . ." Going to what? Of course, she doesn't know. This Anne is a people person. Why would she wish to live in a world without all the entertaining characters with whom she is imprisoned?

Not only did the Hacketts de-emphasize Anne's spiritual and intellectual life, but they also showed scant interest in her moral development, that aspect of the diary that so impressed John Berryman: the conversion of a child into a person. In the play she remains a child, if an erotically awakened one. But what besides time itself, a romance, and some scares about burglaries could have helped her grow? The nightmares to which Berryman attached such importance—the visions of loved ones lost or abandoned to terrible fates, of her friend Lies and her grand-mother—must have been considered way too dark. They've been replaced by a generic bad dream from which Anne wakes up screaming, "No! No! Don't take me!"

And finally there is the line that has come in for so much criticism for its role in distorting our view of Anne and of her diary. "I still believe, in spite of everything, that people are truly good at heart!" It appears not just once but twice in the play. We hear it when Anne tells Peter that the world is "going through a phase, the way I was with Mother. It'll pass, maybe not for hundreds of years, but some day . . . I still believe, in spite of everything . . ." And it is repeated in the final scene, when Otto has returned to the deserted annex. After paging through the diary for the inspirational sentiment, which Anne intones in a voice-over, Otto adds, "She puts me to shame." In one espe-cially vituperative letter to Otto, Meyer Levin called that the only accurate line in the drama.

Even as the principals involved in the Broadway production battled over Anne's Jewishness, even as Meyer Levin claimed to speak for her as a fellow writer, there was no one to fight for an accurate representation of Anne's brilliance and her gifts as an artist. But why would anyone, really? She was only a girl who kept a diary for the last two years of her life.

NONE of this seemed to bother the critics who helped make the play an instant success. In the *New York Times*, Brooks Atkinson wrote that "they have made a lovely, tender drama out of 'The Diary of Anne Frank'. . . they have treated it with admiration and respect . . . Out of the truth of a human being has come a delicate, rueful, moving drama."

Ten days later, again in the *New York Times*, Atkinson re-thought—and heightened—the praise he'd already given the play. "There is only one way to account for the soft radiance of 'The Diary of Anne Frank.' The play and the performance are inspired. At rare intervals along Broadway, something happens that puts the theater on its mettle and animates everyone into doing a little more than he is capable of doing. A dream of impossible perfection drives everyone into lifting himself up by his bootstraps. Something in 'The Diary of Anne Frank' has had that happy effect."

Newsweek's summary lauded "the punch of plain, poignant truth." A positive but peculiar notice in the *New Yorker* contained the following passage: "I can think of no criticism of anything the authors have done, except possibly a tendency toward the end to make their adolescent heroine just a shade more consciously literary and firmly inspirational than either her age or her indicated character would appear to warrant. I have not read the book, however, and it may well be that the quotations that bothered me were taken from it verbatim, since, as I am well aware, most young ladies have their flowery moments. The fault, in any case, is slight, and I think the play on the whole is magnificent."

Only a few critics would remark on what had been lost in the course of Anne's coronation as a Broadway princess. Writing in *Commentary*, Algene Ballif noted that, "If the diary of Anne Frank is remarkable for any one thing, it is for the way in which she is able to command our deepest seriousness about everything she is going through—the way she makes us forget she is an adolescent and makes us wish that this way of experiencing life were not so soon lost by some of us, and much sooner found by most of us. Ironically for her, the Anne Frank on Broadway cannot command our seriousness, for all Anne's true seriousness—her honesty, intelligence, and inner strength—has been left out of the script. . . . If we in America cannot present her with the respect and integrity and seriousness she deserves, then I think we should not try to present her at all. Not all adolescents, even in America, are the absurd young animals we know from stage and screen. . . . Anne Frank was not the American adolescent, as Hackett and Goodrich would have us believe. She was an unaffected young girl, uniquely alive, and self-aware—experiencing more, and in a better way perhaps, than most of us do in a lifetime."

A year after its Broadway debut, the play opened in Germany, where critic Kenneth Tynan observed this response at the end of the performance. "The house lights went up on an audience that sat drained and ashen, some staring straight ahead, others staring at the ground, for a full half-minute. Then as if awakening from a nightmare they rose and filed out in total silence, not looking at each other, avoiding even the customary blinks of recognition with which friend greets friend. There was no applause, and there were no curtain calls." A rather less wholehearted reaction was recorded by Theodor Adorno, who reported that, after seeing the drama, one German woman said, "Yes, but *that* girl at least should have been allowed to live."

Sales of the book spiked in Germany and throughout Europe in the wake of the play's popularity. Perhaps the Anne of the diary was, despite the cuts and edits that had been made for the German edition, still too complicated, too Jewish—and too angry—for the Germans to embrace. But the conciliatory, hopeful, "universal" and dumbed-down Anne made that embrace possible, and Anne quickly became an object of devotion. A memorial plaque now marks the house in Frankfurt where she lived as a small child, and, in 1957, two thousand young Germans made a quasireligious pilgrimage to leave flowers on the mass grave at Bergen-Belsen, where Anne is believed to be buried.

Performed in schools, community centers, and summer theaters, the play has acquainted countless audiences with

Anne Frank and her tragic plight, and has helped steer thousands of readers back to the diary; to a lesser extent, this is also true of the Hollywood film based on the drama. Less happily, the play has, for many people, *become* the diary. In classrooms throughout this country and the world, it is taught to students for whom the flighty, mindless stage Anne becomes the *only* Anne Frank.

And yet there is no denying the effect that the play has on its audiences. I remember seeing the original Broadway production, which meant that I had to have been between eight and ten years old. I was already a passionate fan of the diary, which is how I know how young I was when I first read it. I remember watching the play, feeling that Anne's diary had been brought to life and being so moved that I wept. I remember that the audience was weeping, and that I felt (though I would not have known how to express it then) that a private and personal experience had become a communal one. I could not have been more grateful to have found a theater full of people who shared my admiration and pity for this remarkable girl, and my passion, however childish, for her work.

In 1997, Anne Frank returned to Broadway, in a new adaptation by Wendy Kesselman. Approached by producers Amy Nederlander and David Stone and by director James Lapine, Kesselman undertook the commission because, she says, "I wanted to restore the truth" to the way that Anne had been portrayed onstage. By then, the publication of the 1995 *Definitive Edition* had brought the diary new attention, and those involved in the project wanted to provide a more nuanced and complete picture of Anne's character and achievement.

Kesselman's initial idea was that her version would closely follow Goodrich and Hackett's, with only the "frame"—

the prologue and epilogue in which Otto revisits the attic—eliminated. But as she reread the diary, she decided that more needed to be done, a decision complicated by the fact that the copyright specified that no more than 10 percent of the original 1955 play could be altered.

In fact, Kesselman's adaptation is more faithful to the diary than its predecessor. Anne's voice, her intelligence, and her spirit come through more clearly, and there are longer passages read verbatim from her journal. Anne's references to her physicality—and to her memory of touching another girl's breasts—have been restored. This time, we hear the radio speech by Minister Bolkestein that inspired Anne to think that her diary might be published and to hope that she might grow up to become a writer. The historical and religious contexts have been clarified, as has the threat of what being arrested will mean to the annex residents. We listen to the voice of SS leader Rauter ordering that the Netherlands be cleansed of Jews. (The fact that only a small percentage of Dutch Jews survived was, claims Kesselman, a "revelation" that changed, for her, the popularly held notion that the entire Dutch population was either hiding Jews or working for the Resistance.)

The minor characters are more rounded and more persuasive, and Mrs. Van Daan is given a touching speech about how she fell in love with her husband. In an early public reading of Kesselman's version, Linda Lavin, who played Mrs. Van Daan, was moved to tears the first time she read the passage.

"She doesn't do that very often," an agent is said to have dryly commented.

Anne's belief in the goodness of the human heart has been retained, but returned to what Cynthia Ozick termed its "bed of thorns." The lines about the world being transformed into a wilderness and the suffering of millions are the last we hear from Anne in the play, which ends with Otto informing the audience of how the others perished and how Hanneli saw Anne, naked, her head shaved, ridden with lice, shortly before she died of typhus. Unlike Goodrich-Hackett's, Kesselman's adaptation makes it difficult for the audience to remain in doubt about what happened to Anne.

But finally a play is only a script, a blueprint, and much depends on the quality of the production. In the spring of 2007, a staging of Kesselman's version, directed by Tina Landau at Chicago's Steppenwolf Theater, seems to have maximized its potential.

Responses to the 1997 Broadway production were more mixed. *New York Times* theater critic Ben Brantley was quite taken by the leading actress: "To see Natalie Portman on the stage of the Music Box Theater is to understand what Proust meant when he spoke of girls in flower. Ms. Portman, a film actress making her Broadway debut, is only sixteen, and despite her precocious resume, she gives off a pure rosebud freshness that can't be faked. There is ineffable grace in her awkwardness, and her very skin seems to glow with the promise of miraculous transformation."

Others were less convinced. Writing in *Commentary*, Molly Magid Hoagland noted, "Despite the changes, this is still the sentimental play about a luminous, flirtatious, idealistic Anne Frank that made the critics swoon 40 years ago." In many cases, the critics' disappointment focused on Portman, who appears to have taken too literally the stage directions to run and jump, to fling her arms around her father's neck and prance about. Also in the *New York Times*, Vincent Canby called Portman's performance "earnestly artificial, having been directed to behave in a fashion that might have embarrassed even Sandra Dee's Gidget. Ms. Portman seems never to walk if she could skip; when she lies on the floor, tummy down, heels up, writing in her beloved diary, her little feet are forever kicking back and forth like a 4-year-old girl's. The girl we see has no relation to the thoughts she speaks, either in person or as prerecorded narration." And Portman's own statement, in an interview, that the play "is funny, it's hopeful, and she's a happy person" seems to have been at least partly what raised Cynthia Ozick's psychic temperature to the boiling point at which she wrote "Who Owns Anne Frank?"

For Ozick, Otto Frank is to blame for being "complicit in this shallowly upbeat view," for emphasizing Anne's idealism and her spirit and "almost never calling attention to how and why that idealism and spirit were smothered, and unfailingly generalizing the sources of hatred." To which the reader can only respond by wishing that the sources of hatred were not as generalized as indeed they are. If only the perpetrators and the victims of prejudice were forever limited to Germans and Jews.

Once more, Anne's diary, and the circumstances surrounding it, have given rise to a paradox. Perhaps Otto Frank was right to doubt the wisdom of dramatizing his daughter's work. Perhaps he should have listened to his instincts and resisted the lure of money, fame, and—more important from Otto's point of view—a greatly expanded audience for Anne's book. And yet that audience was, to some extent, generated by the play and the film. We cannot estimate how many readers the play has created for the diary, how many people would never have sought out the book had they not seen the drama first, how many students would never have heard of the diary had the theatrical versions not brought it wider acclaim. In fact, only after the play and the film appeared did the diary begin to be widely adopted as a classroom text.

Though not everyone would agree, one could argue that, in this case, the end result justified the means. Regardless of how the play and the film distorted Anne Frank into a bubble-headed messenger of redemption, regardless of how she was stripped of her religion and ethnicity, robbed of her genius, removed from history and recast as a ditsy teen, the play and the film steered millions of readers back to the diary, which would always remain the diary, no matter how it was misrepresented. As Molly Magid Hoagland wisely pointed out, "There is no need to rely on Broadway, or any intermediary, for a true sense of the brilliant bundle of contradictions that was Anne Frank. Anyone who has a mind to can still turn to the work that Miep Gies rescued and that Otto Frank, despite misgivings, and to his everlasting credit, brought into the light of day. In its pages, in whatever edition, his daughter has always spoken for herself."

Two years after the play's Broadway run, Natalie Portman wrote, in *Time* magazine, about the difference those two years had made in her reading of the diary, a change that one can't help wishing had occurred *before* she took on the role. "At 16, when I portrayed Anne on Broadway, it was her flaws—vanity, overexcitability, and quickness to fight—that interested me the most. And now, upon my most recent perusal just weeks before my 18th birthday, I am struck most strongly by her introspection, solitude, perfect self-awareness and sense of purpose . . . The beauty and truth of her words have transcended the limits placed upon her life by the darkness of human nature."

EIGHT

The Film

AMONG THE TOUCHING ASPECTS OF ANNE FRANK'S ROOM in the secret annex is how much it seems like, and how much it will always remain, the bedroom of a teenage girl. Mostly what freezes it in time and attaches it to a particular stage of life are the movie-star photos, reminders of that longing to be surrounded by celebrity idols whose head shots are, to an adolescent, the height of interior decor. As long as Anne had Greta Garbo on her wall, Hollywood was as near as an attic with blacked-out windows, hidden above the Prinsengracht in the middle of a war.

Anne was a passionate fan of the film magazines that Viktor Kugler brought her, and Hollywood seems to have been very much on her mind. In a diary entry that she cut in her revisions, she imagines going to Switzerland, where a film is being made of her skating with her cousin Bernd. She writes a treatment of the film, which will be in three parts. The first will show Anne skating in a fancy costume; the second will focus on Anne at school, surrounded by other kids; the third will prominently feature Anne's new wardrobe.

One of the stories in *Tales from the Secret Annex*, "Delusions of Stardom," is subtitled "My answer to Mrs. Van Pels, who's forever asking me why I don't want to be a movie star." Dated December 24, 1943, it begins, "I was seventeen, a pretty young girl with curly black hair, mischievous eyes and . . . lots of ideals and illusions. I was sure that someday, somehow, my name would be on everyone's lips, my picture in many a starry-eyed teenager's photo album." The narrator, a Miss Anne Franklin, writes to three movie-star sisters, the Lanes, who write back, inviting her to visit them in Hollywood.

There, "where the three famous stars did more to help their mother than an ordinary teenager like me had ever done at home," Anne Franklin is hired to model for a manufacturer of tennis rackets. But the work is harder than she anticipated. "I had to change clothes continually, stand here, sit there, keep a smile plastered on my face, parade up and down, change again, look angelic and redo my makeup for the umpteenth time." After four days of this, Anne's paleness and general exhaustion convince her hostess that she should quit her job, for which Anne is grateful. "After that I was free to enjoy the rest of my unforgettable vacation, and now that I had seen the life of the stars up close, I was cured once and for all of my delusions of fame."

In October 1942, Anne, who had apparently not yet been cured of her dreams of (or at least her ambivalence about) movie stardom, pasted the photo of herself in her diary, the 1939 portrait that she hoped might improve her chance of getting to Hollywood. In her round, childish print, she spells Hollywood with one *l*. In the same entry, she writes that she had put more film stars up in her room, this time with photo corners, so that she could take them down when she tired of them.

Ironically, the photo did improve her chances of getting to Hollywood, though not in a way that anyone could have predicted, and again at a cost that no one would willingly have wanted to pay.

~

IN 1956, Samuel Goldwyn expressed interest in producing a film of Anne's diary, which William Wyler would direct. But when Otto Frank insisted on retaining script approval, Goldwyn withdrew, a decision he later regretted. Otto signed a contract with 20th Century Fox to turn *The Diary of Anne Frank* into a film with a three-million-dollar budget. It would be adapted from the Broadway play and would also be written by Frances Goodrich and Albert Hackett.

The Hacketts should have known better. The couple's problems, which began almost instantly upon signing the new contract, this time included an emotionally draining ten-day visit from Otto and Fritzi, and a feud with Joseph Schildkraut, who felt that his stage role was being diminished in the film.

George Stevens, whose work included *I Remember Mama* and *Gunga Din*, was an obvious choice to direct. He'd won an Oscar for his last picture, *A Place in the Sun*, based on a Theodore Dreiser novel. He was a serious director who could nonetheless fill seats.

As a lieutenant colonel in the Army Signal Corps, Stevens had headed the combat motion-picture unit, whose members included seasoned Hollywood cameramen, among them William Mellor, who would photograph *The Diary of Anne Frank*. The so-called "Stevens irregulars" not only filmed the Normandy invasion (which provides a dramatic moment in the movie of the diary) but also the liberation of Dachau. His footage was used as evidence at the Nuremberg trials and, today, plays continually on a video monitor that visitors see upon entering the exhibition space of the United States Holocaust Memorial Museum in Washington, D.C. The best-known image from the film is of two boys in their early teens, newly freed prisoners walking down the cobblestone path of the camp. One of them has his arm slung over the other's shoulders, a Jewish Huck Finn and Tom Sawyer in striped uniforms.

Stevens's most recent films, *Giant* and *Shane*, had proved he could work with young actors. *Giant* had featured James Dean, whose restless brand of angst would serve as a template for the roiling adolescent emotions in the secret annex. But *Giant* had only done moderately well; ideally, *The Diary of Anne Frank* would be Stevens's ticket back into the mainstream.

Another hope was that the Broadway cast would repeat their roles, but it soon became clear that Susan Strasberg would not play Anne on screen. The rumor was that she was involved in a distracting affair with Richard Burton; also, she had developed a reputation for being difficult to work with.

Stevens thought next of Audrey Hepburn. Not only had she been born in the same year as Anne Frank, but she was half Dutch and had spent the war in Holland, stranded throughout the occupation with her mother, a Dutch baroness. But that was part of the reason *why* Hepburn turned down the role. Anne's story, she said, would revive too many painful memories. And her age was a problem. She, at least, understood that, at twenty-eight, she would find it hard to play a thirteen-year-old. Besides, she had already agreed to play Rima, the Amazonian bird girl, in the film of *Green Mansions*.

Otto Frank traveled to visit Hepburn and to convince her to change her mind. Otto and Fritzi, Hepburn, and her husband Mel Ferrer spent a day at the tranquil Swiss villa where the actress went to escape the pressures of being a star whose first major appearance, in *Roman Holiday*, had won an Oscar. The Franks stayed through lunch and dinner. But despite the persuasive case that Otto must have made, Hepburn declined. She and Otto stayed friends, and Hepburn, who put her fame behind causes, including campaigns against world hunger and for children's rights, would become an active supporter of the Anne Frank Foundation.

When Stevens's second choice for Anne—Natalie Wood— also passed, a casting call went out for an unknown to play the starring role. The most recent, heavily publicized nationwide search had found eighteen thousand young candidates seeking the title role in Otto Preminger's 1957 *Joan of Arc*. Now the hunt for a newcomer who could play Anne would be appropriately international.

In a newsreel-style promo piece about the making of the film, George Stevens explains that six months had already been spent casting the part of Anne.

"How many little girls have you talked to?" asks the reporter.

Six thousand, of which they'd met half, after which they'd reduced the list to "a hundred interesting possibilities." Auditions were being held in France and Holland. The casting of a Dutch actress would not create a language problem, because "so much good English is spoken in Amsterdam. Many of the Dutch girls that we have heard from have written in very good English, and they wrote these letters themselves."

The main thing was to find a fresh face. Asked if it was true that he didn't want a professional actress, Stevens replied, "We *do* want an actress that hasn't found that secret out yet about herself." He hoped this would not only be the girl's first role, but her *only* role, so that she would be forever associated with the part "and perhaps not others." She need not have a "facsimile resemblance" to Anne. More essential was spirit and "the flavor of Anne Frank in appearance."

They didn't want another Shirley Temple, said Stevens, but someone *like* Shirley Temple in that "she must have charm, she must draw an audience to her, she must draw an audience's affection and its sense of protection." They were looking for a young girl, ideally around thirteen or fourteen. "She could play the younger part of the girl, and then when she puts on clothes, she will do what we often see in children that we know. When they wear their party dress and go out for the first time on a date, we see the youngster through the party dress."

Stevens had visited Amsterdam and talked to Mr. Frank, "an extraordinarily fine gentleman and a survivor of this misfortune." There were plans to do some filming on location in Holland, and when word got out that a Dutch girl might be picked, thousands of letters poured in; around seventy girls had been chosen to audition. The winner, a young, half-Jewish dancer, was ultimately rejected in favor of Millie Perkins, a New Jersey–born Audrey Hepburn lookalike. A former Junior Miss model who had appeared on the cover of *Seventeen*, Perkins was discovered for the part, Lana Turner style, having a snack in a coffee shop with her sister.

For her screen test, she told a story about going to the theater and being terribly annoyed by the people in front of her, a woman who threw her heavy fur coat on top of a little old lady, and a drunk who woke from snoring to guffaw at a serious drama. Aside from her physical resemblance to Hepburn, Millie Perkins's most striking qualities are a brittle perkiness and a highly mannered affect. What she seemed to share with Anne was a mixture of confidence and terror, but a different confidence, and very different terrors.

Perkins would go on to star, opposite Elvis, in *Wild in the Country*, then disappear from the screen to return, decades later, in more "mature" roles. She was briefly married to Dean Stockwell, and, in 1985, played the Virgin Mary in a TV miniseries.

In the 2001 documentary about the filming of the diary, *Echoes from the Past*, Perkins recalls Otto Frank's visit to the set.

"He approved of me and believed in me," she says. As tears come to her eyes, she falls silent and taps her nose, rapidly and repeatedly, then says, "You see I did care."

IN banner headlines, the trailer for the 1959 film promised its audience that "no greater suspense story has ever been told than . . . 20th Century Fox's masterful production of *The Diary of Anne Frank*! Here is the thrill of her first kiss! Here is the wonder of her youth! The excitement of her first love! The miracle of her laughter!" These promises are delivered on by the film itself, a psychological thriller in which the erotic tension leading to a first kiss races against the heroine's inevitable capture by the Gestapo. Presumably, the final cut incorporated the audience responses from test screenings of the film; comment cards (preserved in the Anne Frank archive) asked viewers which scenes and actors they liked most, if any elements of the story were confusing or unclear, if they would recommend the picture to friends. One audience, in San Francisco, objected to an ending in which Anne was shown in a concentration camp, and the closing scene was recut so that Anne was given another chance to proclaim her faith in human goodness.

Otto may have told Meyer Levin that the diary was not a war book, but George Stevens understood that war could keep the action moving. Faced by the problem of how to inject suspense into an essentially static story better suited to the stage, Stevens ramped up the danger outside the annex with footage of prisoners in striped uniforms and the sound of Nazi jackboots hitting the cobblestone streets. The merriment of the Hanukkah party is ended by the menacing hee-haw of Gestapo sirens. There are air raids, bombings, near misses. Dust and fragments of ceiling rain down on the cowering residents. The burglaries start earlier in the arc of the plot and take up considerable screen time. Anne's dream of her starving, suffering friend Lies, altered in the play to an abstract nightmare from which Anne wakes in terror, has become a "dream sequence" in which a girl's tormented face emerges from a background of female prisoners in striped suits.

The film puts Hitler, curiously absent from the Goodrich-Hackett stage drama, back into the picture. His demented voice squawks from the contraband radio around which the annex residents cluster. Stevens compelled his actors to watch the footage he shot in Dachau, and used a recording of crowds shouting, "Heil, Hitler!" to evoke anxiety and fear.

Despite Stevens's efforts to immerse his cast in recent European history, the film seems even more "universal" than the play—that is, less about Jews. The script was sent for approval to the Jewish Advisory Council, an organization formed to monitor how Jews were portrayed on the big screen. Its director, John Stone, not only praised the screenplay but wrote that he preferred it to the play: "You have given the story an even more 'universal' meaning and appeal. It could very easily have been an outdated Jewish tragedy by less creative or more emotional handling—even a Jewish 'Wailing Wall,' and hence regarded as mere propaganda."

After the more disturbing scenes were filmed, Stevens played a loud recording of "The Purple People Eater" to dispel the tension and loosen everyone up. During the shoot, which lasted almost six months, the actors were subjected to a range of physical discomforts intended to re-create the miseries that the annex residents endured. The set was overheated for the summer scenes and excessively air conditioned when the action shifted to winter. Shelley Winters was required to gain fifteen pounds for her portrayal of Petronella van Daan, and then to lose the weight, together with her initially elaborate coiffure.

Unlike the play, the film can step outside the annex, to the pretty streets of Amsterdam and zoom up to its placid sky, where the movie begins and ends to the swells and dips of Alfred Newman's lush score. Yet Stevens was determined to convey the claustrophobia of life in hiding. This posed a challenge because the studio insisted that the production utilize its new CinemaScope technology, which had been developed in the hope that the wide screen would lure audiences back from the tiny windows of their brand-new TVs.

Stevens's solution was to have the set (the interiors were filmed on a sound stage) built vertically so that the camera could pan upward from the office to the garret, catching the annex residents as they wait, frozen with fear, at the top of the steps, or stand amid attic beams, as if they are in a tree house. Heavy supports were constructed and moved as necessary to narrowly frame the shots and counteract the wide Cinema-Scopic panorama.

In close-up, every pore and imperfection becomes gigantic, and the differences among the actors—who appear to have traveled from different countries to act in different movies— are likewise magnified. Puzzling variations in manners, accent, and affect divide the German-born, aristocratic Franks and the Van Daans, who sound as if they have arrived in Amsterdam via the Bronx. Richard Beymer, who had mostly done TV roles and would go on to star as Tony in the film of *West Side Story*, plays Peter van Daan as a rebellious American teen walled up in a Dutch attic.

Millie Perkins appears to have been directed to play Audrey Hepburn playing Anne Frank. Coy and frisky, she pouts, makes faces, and gives little indication of Anne's intelligence and heart. In her memoir, Shelley Winters writes (mistakenly) that Anne knew nothing about the Holocaust. Had she missed the diary entry in which Anne wrote, "If it's that bad in Holland, what must it be like in those faraway and uncivilized places where the Germans are sending them? We assume that most of them are being murdered. The English radio says they're being gassed. Perhaps that's the quickest way to die."

In the documentary about the making of the film, Millie Perkins admits, "I didn't understand the area of the Gestapo and the Nazis." Though she speaks in the Anglo-actressy diction that was Audrey Hepburn's default accent, her voice has an American twang, a cultivated nasality, and at moments she stretches her vowels like the other, more patrician Hepburn, Katharine.

Millie Perkins's struggle underlines the obvious problems of casting an actress to play a character who ages between thirteen and fifteen—a time span during which a girl may feel, and behave, as if she is becoming another person. For all Stevens's talk about the miracle that could be worked by dressing up a little girl and letting the audience glimpse the youngster through the party dress, the result was quite different. An eighteen-year-old dressed up in a child's frock looks like an eighteen-year-old dressed up in a child's frock. It's disorienting and vaguely upsetting to see Anne clinging to her father or sitting with her arms wrapped around his neck and her head on his shoulder; the problem is that she looks like an adult. To accept this Anne as thirteen requires a nearly impossible suspension of disbelief. It's easy to understand how an audience might be surprised to learn that they had been watching a true story.

What was Stevens thinking? He was making a serious movie, coming as close as Hollywood would let him come to European *auteur*dom. Shot in black and white and dimly lit, the diary film was *art*. Besides, the American people needed to know what Jews had suffered during the war. What *people* had suffered.

The documentary *Echoes from the Past* emphasizes the diary's *universality*, the need to make Anne's story "accessible to people all over the world." Stevens "didn't want the audience to think it happened only to Jewish people." According to the documentary, "Stevens's strategy of making the film accessible worldwide paid off" in the form of eight Oscar nominations. Neither a commercial hit nor an unmixed critical success, it won in two categories: Best Cinematography and Best Supporting Actress.

Later, to fulfill a vow she made Otto Frank during his visit to the set, Shelley Winters proudly donated her Oscar to the Anne Frank House Museum, where it is currently displayed in a small case outside the cafeteria. Winters was even prouder of persuading Stevens to restore a section of dialogue that was almost omitted. In the play, Peter wants to burn his Star of David, and Anne suggests that the yellow star is a badge of pride. Realizing that this exchange was missing from the film, Winters interrupted the shoot, angering Stevens, who was eventually mollified into a rewrite.

"I watched him reshoot this scene in which Dick starts to burn his Jewish star, and Anne Frank whispers to him, 'Don't do that. After all, it's David's star—the shape of the shield of his victorious army.' This little moment in the film is extraordinary —the two terrified Jewish children, who are hiding from the Nazis, remembering their heritage of the powerful King David. I will always be proud that I had the courage to stop the filming and see that that moment was restored to the film."

As in the play, Anne's question about why Jews have been singled out to suffer has been changed so that she asks why *people* have had to suffer, first one race, then another. Perhaps because Millie Perkins is so unsure, it's a clumsy moment that nearly brings the film to a stop. You feel as if the actress can't get beyond some problem with this line, and after a beat of hesitation, she sounds as if she's faking it, or lying.

The ending of the movie is also problematic. Anne leans against Peter as they gaze out the attic window. "Some day when we get outside again . . . ," Anne says as the police sirens get louder. Cut to worried adults downstairs, also hearing the sirens, cut back to the lovers staring up at the sky, cut to a truck rumbling down the street. The syrupy music surges under the screech of brakes. The lovers meet in a passionate embrace— "Here is the thrill of her first kiss!"—intensified by the fact that the Gestapo has arrived. The front doorbell rings, Mr. Van Daan faints, there's the sound of crashing, of shouting in German, more crashing, someone's breaking down the door. And the music gets louder. The secret annex residents form a tableau of nobility—eight brave, resigned statues awaiting the inevitable. Peter comes up behind Anne and rests one hand on her shoulder.

"For the past two years," declares Otto Frank, "we have lived in fear. Now we can live in hope." Hope for what, exactly? Already the scene has shifted to a close-up of the diary, and we hear Anne reading, in voice-over, a passage that does not appear in the diary, an entry she would never have written, even if it had been possible for her to write anything at that dreadful moment.

"And so it seems our stay here is over. They've given us just a moment to get our things. We can each take a bag, whatever will hold our clothing, nothing else. So, dear diary, that means I must leave you behind. Good-bye for a while . . . Please, please, anyone, if you should find this diary, please keep it safe for me, because someday I hope . . ."

And so it seems our stay here is over. Our stay here?

Returned from the war, Otto Frank enters the attic with the two helpers, who explain why they were absent on the day the annex residents were taken away. In reality, as we know, they were in the office, and Kleiman and Kugler were arrested along with the Jews; presumably, their presence at that critical moment was edited out to simplify the scene and heighten its dramatic impact. Otto tells his Dutch friends about the camps, about his journey home, about looking for his family among so many others searching for loved ones.

"But Anne . . . I still hoped. Yesterday I was in Rotterdam, I met a woman there, she'd been in Belsen with Anne. I know now. . . ." The music swells again, the camera zooms up for a wide shot of clouds and swooping seagulls, and we hear Anne repeating that, in spite of everything, she still believes that people are good at heart.

She sounds like an American girl. And why not? It's an American movie. We're the cavalry that rides over the hill. In this case the cavalry did its best, but its best wasn't good enough to save Anne. D-day is a major event, and the film utilizes Stevens's footage of the invasion. When his actors failed to respond with sufficient excitement to the news of the American landing on the Normandy beaches, George Stevens played them "The Star-Spangled Banner."

You can watch Stevens's D-day footage on YouTube. The distorted color of the degraded film stock gives it an otherworldly beauty. There's a particularly lovely shot of battleships, silhouetted against the horizon, floating beneath a sky dotted with surveillance balloons. A link takes you to a clip identified only as "Auschwitz Liberation. Rare Russian footage." It's narrated in German and seems to have been made for German TV. A small band of Russian soldiers are running across the fields, falling and stumbling in the deep white snow drifts. Next we see the newly freed prisoners, one by one, men with faces of great strangeness and striking individuality. Then come images of mass graves, corpses in the snow, prisoners of all ages, including children.

The harrowing film reminds you of what was forgotten in the haste to make Anne's diary a lucrative, popular, and morally improving commodity. The camps, the prisoners, and the innocent dead tell the truth beneath the wheeling and dealing of the Broadway and Hollywood productions, beneath the drafts, the rewrites, the lawsuits and disappointments, beneath the simultaneously innocent and cynical American story that ended with a fashion model explaining that despite everything, she still believes that people are good at heart.

NINE

Denial

IN THE SUMMER OF 1998, HELEN CHENOWETH, THEN
the controversial right-wing-Republican congressional representative from Idaho, a strong opponent of gun control and of
environmental protections, was obliged to dissociate herself
from a political consultant named Robert Boatman, who had
produced several video ads for Chenoweth's campaign.

Three years before, the Anne Frank Foundation had sent
its traveling exhibition to Boise. Inspired by the program,
local residents had launched a drive to fund the construction
of a human-rights educational park. Named in honor of Anne
Frank, the park would be located on the bank of the Boise River.
The planners announced that they had already raised almost
$400,000.

This announcement triggered something in Mr. Boatman,
who wrote a letter to the editors of the *Idaho Statesman* that
began, "When a 50-year-old snapshot of a sickly teenager from
halfway around the world appears on the front page of the *States-*

man, you know someone's political agenda is stirring. The perpetuation of the Anne Frank myth by gullible and guilt-ridden crybabies is a slander of truth and a slap in the face of history." The letter went on to claim that Otto Frank had "discovered" and "typed up" the book and subsequently made millions and in the process become "the darling of leftist terrorist groups like the Jewish Defense League and the Wiesenthal Center."

After Representative Chenoweth fired Mr. Boatman, he faded into a shadowy niche of the far-right-wing free-speech pantheon. He is the author of several books on high-speed revolvers, including one on how to "customize" your Glock—a euphemism for turning a handgun into an automatic weapon.

Helen Chenoweth remained in office until 2001.

In the 1960s and '70s, a movement formed, and spread with alarming rapidity, dedicated to championing and publicizing its members' conviction that the Holocaust had never occurred, that the Nazis had never built or used gas chambers and crematoria, and that the number of Jewish World War II dead had been wildly exaggerated. The Nazis themselves had prepared the groundwork for these specious claims, destroying the evidence of the methods used at the extermination camps and employing euphemisms such as "resettlement" and "relocation" for deportation and mass murder.

Aided by neo-Nazis, gathered under the banner of organizations that have included the Institute for Historical Review and the Committee for Open Debate on the Holocaust, these so-called Holocaust revisionists have challenged the so-called exterminationist theorists by placing ads in newspapers and establishing Web sites. Despite the fact that Holocaust denial is illegal in many countries, it has proliferated, drawing some of its most active supporters from the former Soviet Union. It has also gained currency in the Muslim world, where its most visible proponent is Iranian president Mahmoud Ahmadinejad, who has stated that the Holocaust is a Zionist myth, and who, in 2006, assembled, in Tehran, an International Conference to Review the Global Vision of the Holocaust.

If the Holocaust is a fabrication, then it stands to reason—according to the mad logic of "historical revisionism"—that Anne Frank's diary must be a fraud. The first to say so in print was Harald Nielsen, a Danish critic, who, in 1957, published an essay in a Swedish newspaper claiming that the diary was partly the work of an American writer named Meyer Levin. Nielsen's charges were echoed, the next year, by a Norwegian journalist who went further and claimed that the diary was a fake. Surely the most regrettable and unforeseeable consequence of the conflict over the dramatization of Anne's book was that Holocaust deniers would use Levin's lawsuit against Otto Frank as "proof" that the two men had conspired and collaborated to forge a young girl's diary. Why *else* would two Jews sue each other in a New York court for breach of contract and plagiarism?

In 1958, Lothar Stielau, a high school English teacher in Germany and a former Hitler Youth leader, wrote an essay claiming that Anne's journal was sentimental and pornographic, and equated it with the counterfeit diary of Hitler's mistress, Eva Braun. During an official investigation, Stielau took the semantic defense, admitting that instead of the German word for *fake* he should probably have used the word for *seriously altered*. He was defended by a right-wing German political leader, Heinrich Buddeberg, who repeated the accusation that Meyer Levin had been involved in the forgery. Stielau was fired from his job, and both Stielau and Buddeberg were sued by Otto Frank for libel and defamation.

Given that Otto Frank had begun to see himself as Anne's emissary of forgiveness, given that he had declined to prosecute the Nazi officer who arrested his family, or to expose the man who had betrayed their hiding place, his decision to take legal action meant that Stielau and his supporters must have angered and alarmed him. He intuited that Stielau's charges would be repeated and taken up by others, as indeed they were. Unfortunately, the trial failed to serve as a deterrent.

The prosecution argued that the diary needed to be authenticated, since so many vengeful factions were at that point trying to make the Germans look bad. The lawyers mentioned an article in *Der Spiegel* asserting that Anne's work had been heavily edited by Albert Cauvern, one of the friends to whom Otto had first shown the manuscript. Miep and Jan Gies and Bep Voskuijl were brought into court to swear that Anne actually did keep a diary, the same one they had given Otto. Forensic handwriting experts convinced the judge that the diary was authentic.

Still, the case dragged on for three years. Stielau again clarified the point he'd tried to make: he meant the play, not the diary. *The play* was the fraud. In 1961, the lawyers settled. The defendants admitted that the diary was authentic; they apologized and stated that they hadn't meant to offend Otto Frank or the memory of his daughter. Nearly all of Stielau's fine was paid by the German state.

But the verdict in Otto's favor would only have convinced those who already agreed with it. From then on, the books and tracts challenging the authenticity of Anne Frank's diary proliferated like an evil chain letter, each building on the others' demented fantasies as if they were proven truths. Their writers spent pages discussing the "fact" that the forgers were so stupid they wrote Anne's diary in ballpoint pen—ink that was not in use before 1944. They were uninterested in hearing that only six pages, in the entire diary, are *numbered* in ballpoint ink, apparently in Otto Frank's hand. The rest had been written with a fountain pen, and all of it was written by Anne Frank.

In 1967, the *American Mercury* ran an essay by Teressa Hendry, reviving the charges that Meyer Levin wrote the diary, and using, again as proof, the court decision, later reversed, ordering Otto to pay $50,000 in damages to his "race-kin" (in Hendry's phrase) Meyer Levin. The *Mercury* had been founded in the 1920s by H. L. Mencken and George Jean Nathan, and several changes of management later had become the paper of record for the racist intelligentsia.

The most chilling aspect of Hendry's essay is its reasonable, quasi-academic tone. Beginning with a nod to *Uncle Tom's Cabin*, it cites the question that Abraham Lincoln is said to have asked Harriet Beecher Stowe: "So you are the little woman who wrote the book that made this great war?" Hendry riffs briefly on the power of propaganda and on the cleverness with which Communists have used the nonexistent threat of Hitler and Nazism to divert the world's attention from the "live threat" of Stalin and Khrushchev. Then Hendry goes into the details of the Levin-Frank suit, and asks why this case has never been "officially reported." After noting that both Levin and Frank are Jewish, so they cannot be accused of anti-Semitism, no matter what vile lies they tell about Jews, Hendry ends with a plea for truth. "If Mr. Frank used the work of Meyer Levin to present to the world what we have been led to believe is the literary work of his daughter, wholly or in part, then the truth should be exposed . . . To label fiction as fact is never justified nor should it be condoned."

The claim that the diary is a forgery has since been echoed by prominent Holocaust deniers, among them Richard Harwood, author of *Did Six Million Really Die? The Truth at Last*, and David Irving, who also cited the Levin court case as evidence of Otto Frank's complicity in perpetrating the fraud of his daughter's work. When Otto Frank protested, Irving's publishers removed the accusation from his book, *Hitler and His*

Generals, and Irving was ordered to pay damages to the Anne Frank Foundation.

Still more pamphlets were published by a German named Heinz Roth, denouncing the diary as a swindle. The leaflets he distributed at a 1976 Hamburg performance of *The Diary of Anne Frank* aroused the interest of a German prosecutor, who issued an injunction prohibiting Roth from handing out his broadsides. In Roth's defense, his lawyers cited a book that would become a sacred text to those who challenged the diary's authenticity, *The Diary of Anne Frank—Is It Authentic?* by Robert Faurisson, an early champion of the idea that the lying fairy tale of gas chambers and crematoriums had been perpetrated by the Allies and Jews to defame the heroic Nazi party.

Subsequent cases against later pamphleteers were dismissed on technicalities or on the grounds of free speech, a fundamental principle that inspired Noam Chomsky to write the introduction to one of Faurisson's books. Only one journalist, Edgar Geiss, arrested for distributing pamphlets in the courtroom where a colleague's case was being tried, received a criminal sentence. He was given a year in prison for defamation, a judgment he later appealed.

DITLIEB Felderer's 1979 *Anne Frank's Diary, A Hoax* is still among the most repellent attacks on the diary. It is at once tedious and terrifying to attempt to follow the lunatic convictions that keep Felderer (an Austrian Jew who became a Jehovah's Witness, emigrated to Sweden, and was won over to the "revisionist" cause while investigating the Nazi persecution of his fellow Witnesses) raving for page after page, spinning a seamless web of pure, poisonous hate.

Only the like-minded, or those with strong stomachs, could read more than a few lines of his rant, all in the name of history, truth, science, common sense, and inside information. Today, an Internet search of his name directs one to a site headlined "To inform man is not a business but an obligation" and "limited censorship is the root of all terrorism." According to the text, "Ditlieb Felderer's Flyers have come out in the millions in spite of corrupt Politicians Fahrenheit 451ing them over and over. To see the various court records surrounding his censorship trials, Biblical sex photos and discussion, blog, studies of medieval history, controversies, dissent . . .". There is a link to a site that lists Anne Frank and Auschwitz as well as "holocaust-sex" and "pornocaust."

One learns from Felderer's book that the wearing of the six-pointed star was the Jews' own idea, a "fact" proved by a slogan, allegedly coined by a Zionist weekly, exhorting fellow Jews to wear the badge proudly. Apparently, the star was something like a professional or guild badge, or like the lapel pin worn by the French Legion of Honor. We read about Otto's Frankfurt family, "wallowing in wealth," Jews characteristically not satisfied with owning a little piece of Germany and wanting to possess the whole country. Felderer goes into detail to establish the fact that the windows of the secret annex could not have been covered with paper, as the diary claims, that the adult males could not have been heavy smokers without alerting the warehouse staff, that the annex residents ate like kings, though (paradoxically) they could not have cooked without their presence being detected.

The very idea of the helpers spending the night in the annex strikes Felderer as so reckless that the diary entries describing these overnights are in themselves enough to prove that the book is a lie. No sensible Dutch person would save a journal full of anti-German insults in a desk drawer where any German could discover it. And why would Anne have kept her diary in her father's briefcase, where her father could have found it and have read her filthy little secrets—a question that, among other things, betrays a failure to understand the respect for privacy, even the children's, that kept the annex residents civilized and sane.

The worst parts are the dirty bits. Felderer calls the diary, "the first pedophile pornographic work to come out after World War II and sold on the open market. In fact, the descriptions by a teenage girl over her sex affairs may likely be the first child porno ever to come out." He suggests that the "sex portions" may be made up, invented by adult men in order to sell a book that otherwise might have wound up among Otto Frank's private papers. As upsetting as it is to imagine Otto Frank using his daughter "in such a filthy manner," Felderer points out, evil parents regularly prostitute their children, "so why could literary prostitution not be possible?"

"Apparently the 'sexy' portions were too much even for some Jews to stomach, and one of the first, if not the only group, to voice their objections against the diary, were some Orthodox Jews who felt it gave the Jews a bad moral image . . . Whether their objections were based on true moral grounds or for fear that the story was letting the cat out of the bag may be debatable. Talmudic sources are certainly not foreign to perverse sex."

Just when one imagines that things cannot possibly get uglier, Felderer includes a section entitled "the Anal Complex," in which he makes a seeming half turn and argues for the possible *authenticity* of the diary. The only reason the diary *might* be the real thing is "its preoccupation with the anus and excrements, a trait typical of many Jews. Pornography and excretal fantasies have always fascinated many of them and they have therefore also been the greatest exploiters of these things."

He cites Anne's account of the schedule for using the lavatory, as well as her unselfconscious description of the problems that flatulence caused in the airless attic. Among the other anally fixated Jews whom Felderer accuses along with Anne Frank are Sigmund Freud and Charlie Chaplin. Though Chaplin was not, strictly speaking, Jewish, he was constantly scratching his buttocks, and, in one film, he lampooned Hitler, all of which made him Jewish by association. Felderer wonders why no Nobel Prize has yet been awarded for the "science" of anal eroticism.

THESE and other attacks on the diary were, in part, what prompted the Netherlands Institute for War Documentation to engage the State Forensic Science Laboratory to undertake the investigation that produced more than 250 pages of findings proving the diary's authenticity. Experts examined and dated the physical materials from which the diary is made—paper, glue, ink, fibers from the cover—to prove that they were in use before 1944.

The pictures and postcards Anne pasted into her journal were also examined and dated, but most of the study involved an analysis of her handwriting. *The Critical Edition* includes an exhaustive account of characteristics and microcharacteristics, spacing arrangements and interaction points, "the movement component vertical to the writing plane" and changes in pen pressure. The conclusions are clear. Both the block printing of the early pages, and the cursive of the later entries and of the revisions, are those of the same girl. Anne Frank's handwriting changed in ways that fell well within the predictable parameters for a period of two years. The corrections added by other hands at a later date are as minor as they are rare.

Though the attacks on Anne Frank's diary were loathsome, the publication of *The Critical Edition*—which features the find-

ings of the forensic investigators, as well as numerous repro-
ductions of pages from the diary that illustrate and confirm
the committee's findings—is a cause for gratitude. For the first
time, readers could, if they wished, compare the three versions
of the diary and get a sense of how Anne Frank's style developed and what she meant to include in *Het Achterhuis*.

None of this impressed those who believed the diary was
a fraud. *The Critical Edition* failed to persuade Holocaust revisionists who preferred the idea of two old Jews making a fortune by faking the diary and perpetrating Zionist lies about gas
chambers and camps. The diary's critics argued that the variant
drafts were further evidence that it was a counterfeit.

Shortly after the publication of *The Critical Edition*, the *New
York Times* reported that attacks on the diary had increased.
"We never had illusions that this would stop the hoax claims,"
an official of the Anne Frank Foundation told a *Times* reporter.
"But it was surprising the way neo-Nazis in Austria, for example, rejoiced at all this new information and insisted the Dutch
government had offered proof of the falsity of the diary."

Ten years later, at the premiere of the University of Hous-
ton's production of Wendy Kesselman's adaptation of the
Goodrich-Hackett play, members of the National Alliance, a
neo-Nazi group, distributed copies of a leaflet entitled "Anne
Frank Hoax Exposed" in the theater's parking lot. The essay—
purportedly by William Pierce, author of the infamous *Turner
Diaries*, an apocalyptic race-war novel popular among white
supremacists—again repeated the charges that Otto Frank had
forged Anne's journal.

In 2006, the mayor and the police chief of Pretzien, fifty
miles from Berlin, watched a group of young, beer-drinking
neo-Nazis throw Anne Frank's diary into a bonfire, along with
the American flag. When seven of the revelers were tried for

sedition, a defense lawyer claimed that his client had been mis-understood, and that his intention had been to " 'symbolically free himself' from the gloom cast by the Nazi period on 'an evil chapter' in German history." Five of the seven men were con-victed, fined, and sentenced to nine months' probation.

A Google search using the words *Anne Frank Holocaust denial* first turns up pages of legitimate sites about Holocaust denial, then descends into the vortex of bigotry and hate. A more di-rect route is via *Anne Frank Hoax, hoax* being a buzzword and a coded entry into the world of defiant racism.

On Yahoo, there's a list of Anne Frank chat rooms, each with a slogan hinting at what may be found at the end of one thread or another. The group that logs on to explore "The fictional life and times of Anne Frank, the young lover of Herr Adolph Hitler" is closed to new members, and membership is required to join the discussion on "Anne Frank—The Truth"—the truth apparently being that "Mr. Frank betrayed his own family to escape justice." This theme has some currency, attracting yet another group under the rubric, "Anne Frank was betrayed by her own father, what more could you expect from the Jews?" The majority of the chat rooms have more obscene and vio-lent slogans, and share graphic fantasies about Anne's sexual kinks, her enthusiasm for oral sex, her fondness for showing her breasts.

What makes it all the more frightening is that these groups twist every mention of Anne Frank into new evidence that the diary is a fraud. For them, the very idea that a brilliant girl would keep a daily journal of her time in hiding and then go back and revise it because she wanted her book to be published is final, irrefutable proof that the Holocaust never happened.

PART IV

Anne Frank in the Schools

TEN

Teaching the Diary

The *Diary* is many things at one and the same time. It is an amusing, enlightening, and often moving account of the process of adolescence, as Anne describes her thoughts and feelings about herself and the people around her, the world at large, and life in general. It is an accurate record of the way a young girl grows up and matures, in the very special circumstances in which Anne found herself throughout the two years during which she was in hiding. And it is also a vividly terrifying description of what it was like to be a Jew—and in hiding—at a time when the Nazis sought to kill all the Jews of Europe.

—CLIFF'S NOTES ON *The Diary of Anne Frank*

According to a 1996 survey cited on the Anne Frank Museum Web site, 50 percent of American high school students had read *The Diary of Anne Frank* as a classroom assignment. Each year, Anne's diary makes its way into thousands of schools, and onto the desks of teachers who discover that the book most certainly does not, as they say, teach itself. The diary and, more to the point, the circumstances surrounding the composition of the diary, are difficult for students to take in, especially if Anne's story represents their first exposure to the horrors of the Nazis' war against the Jews—which is often the case. Another survey, conducted in 2008, found that only a quarter of American teenagers were able to identify Hitler.

And so it happens that teachers—undercompensated and overburdened by crowded classrooms—must assume yet another challenging task, one that seems essential for their students' historical, literary, and moral education. They must stand in front of a room of bright-faced young people and inform them that not very long ago, the German government and its army cold-bloodedly gassed and brutally murdered millions of Jews, gypsies, homosexuals, Poles, Jehovah's Witnesses, and political opponents, unless—in a very few cases—they managed to hide in caves and haylofts and attics, like the little girl whose book survived, though its young author did not.

Doreen Hazel, a former teacher, offers a course that meets once weekly, for ten weeks, at Manhattan's Anne Frank Center, a nonprofit organization, allied with Amsterdam's Anne Frank Museum, which develops educational programs and hosts exhibitions and workshops in its SoHo loft space. Hazel's class, Bringing Anne Frank to Life in Your Classroom, is open to the public, and teachers can take it for continuing-education credit.

On the afternoon I visited, the class had three students. One was a grade school teacher, the second taught middle school. The third, who was Dutch, was not a teacher—just curious, she explained.

Hazel had loaned the middle school teacher a film about Anne Frank to show her class. When Hazel asked how it had gone, the teacher shook her head and handed back the tape. She reported that, during a scene in which naked female prisoners are being taken to the showers to be gassed, one of her students became completely hysterical and began sobbing with fear that the same thing might happen to her and her mother.

"It was too much for her," the teacher said, shaking her head again. "I stopped it." Her first concern was that the girl might be traumatized; a secondary worry was that, if the student turned out to be inconsolable, her mother might complain to the school. "The women were *naked*," the teacher repeated, widening her eyes and communicating with a glance what the consequences of this could have been for her job. The helplessness, alarm, and sympathy she'd felt was still visible on her face. Watching her, I understood, as I had not before, how *painful* it can be for teachers to address this material, and why they so often cannot bring themselves to do what I had assumed *any* teacher could, and should, do—that is, to simply tell their students the facts of the Nazi genocide. Obviously, these truths constitute an important and necessary lesson, but it was suddenly clear to me how difficult that lesson can be, and why teachers may be so eager to move, as quickly as possible, from the devastating to the constructive, from the historical to the personal.

Faced with the challenge of presenting Anne's diary to a class, even the most tough-minded and confident teacher might take a deep breath and look for some professional or collegial guidance. One essay, "Teaching the Holocaust," by Rebecca Kelch Johnson, published in the *English Journal*, suggests that the Nazi war on the Jews is such an appalling narrative that to teach it as a singular historic event—without emphasizing the importance and value of human rights and the efficacy of nonviolence—would make it seem as if teachers were trying to shock their students with a gory, sensationalistic horror story that would only alienate and depress them:

> *Perhaps teachers hesitate to teach . . . students about an*
> *historical episode which is marked by incomprehensible sav-*
> *agery and abhorrence. The very idea of lecturing about the*
> *fact that between 1933 and 1945 six million Jews along with*
> *others . . . were dehumanized by starvation and subsequently*
> *annihilated by gassing, mass executions, and other methods*
> *employed by Hitler's henchmen is unthinkable. Discussing*
> *the horrors that went on in concentration camps . . . may seem*
> *too barbaric. Yet, an understanding of the Holocaust should*
> *stimulate every Jew and non-Jew's personal understanding of*
> *human rights . . . The adolescent's idealistic nature encour-*
> *ages discussion and study of human rights and justice.*

But even the most positive classroom discussion of human rights eventually comes up against the grim reality of what happened to Anne Frank, and even the most hopeful consideration of the power of nonviolence must address that aspect of human nature that Anne wrote about with honesty and concision: *"There's in people simply an urge to destroy, an urge to kill, to murder and rage, and until all mankind, without exception, undergoes a great change, wars will be waged, everything that has been built up, cultivated, and grown will be destroyed and disfigured, after which mankind will have to begin all over again."* Teachers will inevitably be obliged to face the natural discomfort—particularly strong among the young, and, for some reason, among Americans—with questions for which there are no simple answers, or, worse, no answers at all. Here, the insoluble mystery is that of evil, of the aberrant strain in human nature that fueled the Nazis' efforts to exterminate entire populations.

The bright thread that can be, and frequently has been, teased out of the diary's dark content runs through a lecture delivered in 2003 by Dr. Lesley Shore, then an assistant professor at the University of Toronto. Anne Frank, noted Shore, "champions goodness—and is frequently dismissed for it . . . Anne, like Sophocles's Antigone, chooses to 'join in loving not hating' though, like Freud, she understands the evil that lurks within . . . Anne Frank brings out the best in us. We love her because we know that, could we believe as she did, in the face of terror, if not in the squalor of Bergen-Belsen, we want desperately to believe in the goodness of humankind. This is the power of her legacy. She, alone, dares to admit that she wants to believe that people are basically good at heart."

THE prospective teacher of Anne Frank's diary can find extensive help in the form of books, workbooks, essays, journals, and Web sites. Nearly all of these sources make interesting reading, even the jargon-laden academic essays larded with terms like *emplotment* and *enfigurement*. Predictably, attempts to reduce a historical catastrophe to a series of short-answer questions only succeed in highlighting its irreducibility. One classroom guide, Michelle Keller's *Remembering the Holocaust*, features the following sample test.

> *True or false: The only people killed during the Holocaust were Jewish.*
> *True or false: The Holocaust could never happen again.*
> *True or false: The Holocaust took place during the American Revolution.*
> *True or false: Anne Frank published her diary and made lots of money.*
> *True or false: Two out of every 3 Jewish people in Europe were killed during the Holocaust.*

Obviously, the Holocaust did not occur simultaneously with the American Revolution. But not even the simplest of the other questions is simple. The numbering of the Holocaust dead has caused bitter rifts among historians and political and religious figures. The issue of whether the term "Holocaust" applies only to the murder of the Jews or if it also refers to the Nazis' other victims has been a divisive one. The reference to Anne making "lots of money" taps into a fixation of those who claim the diary is a hoax perpetrated by greedy Jews. How could the statement that "the Holocaust could never happen again" be either true or false, given that genocides have occurred after World War II. *True* would be the correct answer if we argue, as many have, that the Holocaust is a singular event, but the wrong one if we interpret the question more broadly to mean other attempts to destroy a race or religion, tribe or nationality.

A series of comprehension tests from 1993 include questions that any teacher—that anyone, really—would want students to get right. ("Hitler's 'final solution' to the 'Jewish question' was a. extermination, b. deportation, c. relocation.") But other sections may direct the class's attention in odd directions. ("The big disadvantage to keeping cats in the 'Secret Annex 'was the a. litter box smell, b. horrible fights they got into, c. fleas.") Some questions are all but impossible to answer: "Anne's one golden rule was to laugh about everything and to a. not bother about the others, b. not take anything too seriously, c. keep your troubles to yourself."

If the multiple-choice tests seem overly simple, essay suggestions and lesson plans are often bewilderingly abstract. According to *A Guide for Using Anne Frank in the Classroom*, students should prepare for the topics the diary raises by responding to the following statements with a simple "agree" or "disagree."

1. *I want my memory to live on after my death.*
2. *Until mankind undergoes a great change, there will be wars.*
3. *What is done cannot be undone, but one can prevent it from happening again.*

4. *The final forming of one's character lies within one's own hands.*

5. *It is good to always follow one's conscience.*

6. *In spite of everything, people are really good at heart.*

The metaphysical nature of these prompts seems like further evidence of the educator's (and indeed, our own) distress at the horror of Anne's fate. This discomfort often leads teachers to gloss over the pathology that drove Anne into the attic and to focus on the resilience and spirit of the Franks and their neighbors. However understandable, this impulse does Anne and her work a disservice, since, in her case, neither element—the will to survive with the maximum humanity and the will to extinguish with the maximum brutality—makes sense without the other.

The desire to extract an affirmative lesson from Anne's story likewise explains the fact that the Anne who visits the classrooms, and whom the pedagogical literature describes, more closely resembles the Broadway and Hollywood Anne than the Anne we meet in the diary. Much is made by the teaching guides of her optimism and resilience, but there is little acknowledgment of the fact that she was a complicated young artist who died a tragic early death. An article in the *Journal of Adolescent and Adult Literacy*, by Stephanie Jones and Karen Spector, notes that students may actually resist the suggestion that Anne's story is not the sunny narrative they wish to imagine:

> Even when students were explicitly told of her cruel death, they still tended to imagine her in hopeful ways. When students answered a question in their textbook . . . that asked how Anne could have been happy in a concentration camp, Charlotte answered, "Knowing Anne, she was happy in the concentration camps. She didn't have to be quiet anymore; she

*could frolic outside. She could be in nature. She loved nature.
I think this was a welcome relief for her." The basis for Charlotte's version was simply, "Knowing Anne . . ." When Karen
asked Charlotte's classmates if they agreed with her, the room
was filled with lifted arms; some had both hands raised, yet
no one raised a voice or kept an arm down in protest of Charlotte's statement. No one. This is a testament to the powerful
pull of the Americanized Anne Frank.*

One can imagine that Jews who had been in hiding would
have been (however briefly) glad for the relative freedom of
Westerbork. But the idea of Anne frolicking in Auschwitz or
Bergen-Belsen suggests a flaw in the Holocaust units of which
her diary often forms the core. Some responsibility for this may
stem from the cognitive dissonance that must affect teachers
attempting to present (and the student trying to grasp) the
life of Anne Frank as an example of the triumph of the human
spirit. The logical conclusion to that story is not the mass grave
at Bergen-Belsen, and so it must be tempting to proceed as if
Anne's story ended when her diary ends, as if Auschwitz never
existed.

A "Cyberhunt Teacher's Page" on the www.scholastic.com
Web site includes a paragraph headlined "Anne's Journal of
Hope," and instructs teachers to focus on the passage that the
site entitles "On Still Believing," the section in which Anne's
belief in human goodness is tempered by the vision of a catastrophic future. After reading this aloud with their students,
teachers should "ask them to reflect on Anne Frank's words in
their own response journals. What do they think she meant by
wilderness and approaching thunder? What does this passage
tell us about her? How does it make them feel? Encourage students to add poems, drawings, and questions in response."

The scope of the discussion suggested by these questions is

fairly common in the teaching guides, but the mention of "Anne Frank's words" represents a departure. Mostly, the lesson plans encourage the instructor to shift so rapidly from Anne Frank to the Holocaust and from there to the students' own experience of prejudice and discrimination that, more often than not, Anne seems to be absent from school on the days when her diary is being taught.

Time and again, the diary itself is used to encourage students to talk about themselves. In an essay, "Literature as Invitation," Robert Probst describes visiting a classroom outside San José, Costa Rica. Enthralled, the observers watched the talkative students move from a discussion of how people can treat one another as inhumanely as the Nazis did to questions about whether such an event could recur, perhaps nearby—and from there to the then-current fighting in Bosnia and Kosovo. The experience, writes Probst, represents "what literature is or might be in the classroom if we respect its power and respect what it offers."

But even if some of these pedagogical approaches fall short of grappling directly with the beauty—and the terror—of Anne Frank's life and work, there is something admirable about any instructor who teaches *The Diary of Anne Frank* in *any* way. (Except, of course, the way in which one can imagine it being taught by the evil Ditlieb Felderer or by Lothar Stielau, who, in addition to being one of the earliest Holocaust deniers to challenge the authenticity of the diary, was a high school teacher.) Ideally, the diary should be presented with the optimal balance of the literary, the historical, and the personal, engaging the entire class, informing each student of something he or she needs to know and leaving them all resolved to be more empathic human beings.

The question of how the diary is taught in our schools is in some ways similar to that of how it has been represented on stage and screen. Even the most unfocused schoolroom debate may inspire students—as it has audience members—to return to the primary source. Again, the diary remains the diary, and in each class, a few readers will feel a connection to its writer and learn something about the times in which she lived. Even when Anne's journal is used primarily as a springboard for personal confession, students will respond to her voice and her sensibility.

But it's harder to persuade ourselves of this when, in many classrooms, the Goodrich-Hackett drama is taught as if it *were* Anne Frank's diary. The advantages of teaching the play rather than the book are obvious. The drama has been effectively precensored and prevetted for an acceptable balance of the upsetting and the uplifting. The references to menstruation have been removed, and Anne's suffering has been "universalized," thus facilitating the transition from talking about the text to talking about the students' lives. It can be employed by both the English and the drama departments in schools where drama departments still exist. First students study the text, then they perform it.

Yet even this can be transformed into a useful exercise. In her intelligent essay, "Drama for Junior High School: The Diary of Anne Frank," Elizabeth A. Mapes describes assigning students to read both the diary and the play, thus enabling them to discuss the differences between literature and adaptation.

PERHAPS I should explain what *I* might do if I taught high school and were fortunate enough to be part of a school system that allowed teachers to decide how to present a book and how much time they would have to do so. I'm imagining a dream class, perhaps an honors senior English class in an urban public school such as New York's DeWitt Clinton High School, a place I mention because I have visited twice, and both times found the students—local kids and the children of immigrants from all over the world—to be bright, eager, well-informed, and several beats quicker than I was.

I might use, or recommend, as a sourcebook, Hedda Rosner Kopf's sensible and comprehensive *Anne Frank's* The Diary of a Young Girl: *A Student Casebook to Issues, Sources, and Historical Documents*, a volume that contains useful bibliographical and background information, a perceptive exposition of why the diary is so extraordinary, a family history of the Franks, and an account of what happened to the Dutch Jewish children caught in the Holocaust. My ideal class would know all about Hitler, but to make sure, I'd spend a preparatory period discussing the rise of Nazism and the final solution.

An excellent essay, Judith Tydor Baumel's "Teaching the Holocaust through the Diary of Anne Frank," includes a list of diary passages that can be used to explore the subject of the Nazis' crimes against the Jews. ("Let us take the *Diary* and begin charting a blueprint by asking some pertinent questions. First, how is Nazi anti-Jewish policy in occupied Holland expressed through the entries written by the young Anne Frank? In what way does her diary chart Jewish response to Nazi policy?")

Many teaching guides suggest immediately opening up the discussion to other genocides and to the students' experiences of injustice, here or in their home countries. But in my class the main event would be the diary itself. I would assign my students to read it at least twice and to write about how their responses change from one reading to the next. In the classroom I'd ask them to read aloud whole entries. Perhaps the communal potato peeling, or the sausage making, or the break-in downstairs. We'd look at how Anne begins and ends an entry, at which details she chooses and what she omits. How does she make us see her family members and neighbors as complex human beings? How does she help us understand their perilous situation? We'd talk about suspense, honesty, tone, and style; about how the threads of plot and character are interwoven; about how an author can seem to be speaking directly to us. I'd remind them that Anne Frank was a writer, that the seeming artlessness of her style is an artistic achievement, and that her accomplishment is in no way diminished (or, for that matter, affected) by her age or gender. I would point out that she never wanted her work to be called *The Diary of a Young Girl*, but, rather, *Het Achterhuis*. Perhaps we would discuss the ratio between hardship and the speed at which children are forced to grow up.

Doubtless there are teaching guides that propose the approach I've described. But in searching the libraries, bookstores, and the Internet, I haven't found much like it. Rather, I've read a wide range of instructions with a few common themes, mostly concerning prejudice and tolerance. Obviously, there is a place for that in the classroom, especially if we acknowledge that the purpose of education is not merely to fill a student's head with information, however practical and useful, but also to teach them how to live.

Anne Frank's diary is full of lessons that can help students reach their own conclusions about morality and human kindness. But the talent, determination, hard work, and expertise with which a child created a work of art, and the tragic circumstances that forced that creation, are themselves lessons that should not be ignored because they make us unhappy or uneasy. The fact that a girl could write such a book is itself a piece of information, as valuable as any of the improving moral principles that can be extracted from the words that a lonely child, imprisoned in an attic, confided to her imaginary friend.

WHAT makes one feel even more grateful to teachers who take on *The Diary of a Young Girl* is the campaign that has been waged to prevent the diary from being taught at all. Anne Frank's diary is among the most frequently banned or challenged books in American libraries and schools.

In a list compiled by the Online Computer Library Center in 2005, *The Diary of a Young Girl* was number 13 on a list of censored books. The National Coalition Against Censorship reports that in the summer of 2004, "At Fowler High School in Fowler, CO, first-year teacher Sara McCleary was not rehired because she assigned to ninth-grade English students *The Diary of Anne Frank*. After a parent objected to a sexual reference, the School Board terminated her contract and removed the book from classrooms, leaving a single copy in the library."

For some years, these attacks on the book—and on those who taught it—were mostly responses to Anne's reflections on her changing body and her infatuation with Peter. Such passages, it was argued, encouraged an atmosphere of sexual permissiveness inappropriate in a classroom setting—or, presumably, any setting where teenagers were present. A 1982 survey found that the book had been banned in part because "it describes a young girl's physical development too explicitly." These objections unwittingly echo Ditlieb Felderer and others like him who have denounced the diary as a depraved sex book. In fact, one reason the diary has remained so popular with young readers has to do with Anne's forthright and nonhysterical approach to sex, a topic even the most savvy teenagers often find threatening or embarrassing. How unfortunate that the book should be removed from the curriculum because of something it does well—conveying so accurately an adolescent's growing awareness of sexuality in a way that still feels honest and true.

Later, as the mood at district school boards shifted further toward the right, charges against the diary would expand to include Anne's rebelliousness, which was seen as an implicit encouragement of its adolescent readers' lack of respect for authority. More recently still, the list of parental and community objections has expanded to include the very same moral and spiritual values the diary is regularly used to foster.

In December 1983, seven fundamentalist families sued the public school district of Hawkins County, Tennessee, claiming that a textbook taught in their children's classrooms exposed students to values and ideas—secular humanism, liberalism, religious diversity, tolerance—that violated the families' most deeply held beliefs. In the suit, *Mozert v. Hawkins County Board of Education*, they charged that "the use of certain texts violated their right of free exercise of religion and the fundamental right of parents to control the religious and moral instruction of their children."

The case began when one mother, Vicki Frost, found her child's textbook to be rife with references to witchcraft and magic. The opening of *Macbeth* and a selection from the *Wizard of Oz* were particularly offensive. She was further incensed by an excerpt from Anne Frank's diary suggesting that it doesn't matter which God you believe in so long as you believe in *a* god.

The passage was from the Goodrich-Hackett play, not from the diary. Ironically, the controversial lines are taken from the same scene that caused Meyer Levin such grief, the discussion in which Anne tells Peter that she wishes he had a religion, then says that he doesn't have to be orthodox, or believe in heaven and hell. "I just mean some religion . . . it doesn't matter what."

How it hurt poor Levin, and later enraged Cynthia Ozick, the deliberate dejudification of Anne and her family, and by implication the millions of others who died in the Nazi camps. The result was no less distressing to fundamentalist Christians, a situation that the play's creators—so eager to *universalize* the material's appeal—could hardly have predicted. One wonders how the fundamentalist community would have reacted to the diary itself, in which it is clearer that the people in the attic are not Unitarians, but Jews, and that Anne does indeed have a preference about which God she worships. Even as some readers and critics decried the ways in which the diary was, in Ozick's words, "bowdlerized, distorted, transmuted" into a plea for understanding and acceptance, others charged the book (or actually, the dramatic adaptation) with being *too* tolerant.

According to the court documents in the *Mozert v. Hawkins County Board of Education* case, "It is this underlying philosophy that offends the plaintiffs who believe that Jesus Christ is the only means of salvation. Plaintiffs reject for their children any concept of world community, or one-world-government, or human interdependency. They also strongly reject any suggestion, by implication, that all religions are merely different roads to God, finding this an attack on the very essence of the Christian doctrine of salvation."

During the trial, the plaintiffs further charged that the readings fostered rebellion and anarchy, and that both parents and children could face eternal damnation as a result of merely coming into contact with the "evil," "polluted," and "heathen" texts.

The court failed to find that either the textbooks or the cited selections were in violation of the plaintiffs' constitutional rights. Three years later, when the case was appealed, a federal judge ruled that the plaintiffs' children could be permitted to skip the reading classes; the school board was ordered to pay the families over $50,000 in damages.

In 1987, this ruling was reversed by appellate judge Pierce Lively, who made a distinction between reading about other people's beliefs and being forced to adopt them. While the case lingered in the courts, a theater director asked for permission to stage *The Diary of Anne Frank* at a Hawkins County high school. The school superintendent refused, out of concern that the production might further offend the fundamentalist parents.

IN the spring of 2008, I was invited to visit an art class at Bell Academy, a public middle school in Queens, a charter school whose students range from gifted to those with learning disabilities. Taught by Andrea Kantrowitz, an artist, the class met for two hours on Friday afternoons, over a period of seventeen weeks. Its students, who represented the spectrum of kids enrolled at Bell Academy, were each given a paperback copy of *The Diary of a Young Girl* and encouraged to keep a journal into which they could copy their favorite quotes from the diary and write their own responses and stories. Their in-class project—for which they were divided into small groups, working collaboratively—was to produce an anthology of comic strips on themes related to Anne Frank's work or to discussions it had inspired. They also studied passages from two graphic novels, Art Spiegelman's *Maus* and Marjane Satrapi's *Persepolis*.

It was an elective class—self-selected—and the students seemed happy to be there. Every one was clearly paying full attention, even a few solitary kids who seemed detached from the group. The class included whites, blacks, Latinos, South Asians, and Koreans; one pair of boys appeared to be mildly autistic. These last were among the most interesting students, certainly the most devoted to copying quotes and writing in their journals. One of them had begun a tale of time travel, or "time busting," a journey that took its characters from the present day back to the Holocaust.

The session I attended was conducted like a studio art class. Kantrowitz tactfully critiqued the students' cartoon strips and suggested improvements. The plots were allegories about racism, narratives affirming that respect and affection could transcend barriers of color, even of species. One comic strip concerned a giraffe and a donkey playing on the swings on a sunny day in the park when a troublemaking rabbit laughed at them for being friends even though they had different markings; this challenge left the two animals briefly nonplussed, and then more determined than ever to get along.

As I went from group to group, talking to the kids, most of whom had read all or part of the diary, I was struck by the words they used about Anne Frank. *Brave*, said one. *Unselfish*. What amazed one boy was that Anne could still think people were good at heart when she was "all cramped up" in the attic. "It made me think that people are always suffering somewhere," said a girl, "and how lucky we are that we can go to school."

One little girl said that the diary had comforted her, because she was Jewish, and she'd had a really good friend in grade school, and then one day her friend told her that they had to stop being friends; her dad didn't want them hanging out because she was Jewish. The girl beside her said she'd had the same experience with the same girl in grade school. "She wasn't allowed to be friends with us because we're Jewish. Of course, we're not in the Holocaust," she said. "We know that."

"There's still racism," added another girl. "But not here in this class."

I looked around. I thought, she's right. I thought of Mariela Chyrikins and Norbert Hinterleitner. Their jobs here would have been easy. These kids weren't tomorrow's fascists and skinheads. For them, reading the diary was less of a critical intervention than the widening of their circle of acquaintance to include a girl who lived and died long before they were born and who was right about the fact that hope and suffering, compassion and prejudice will be with us forever.

ELEVEN

Bard College, 2007

> The diary is a second kind of Secret Annex, and it is
> where we remain with Anne, hearing her speak to us
> only once every few days and sometimes only for a
> moment because we must keep quiet so as not to let
> anyone know that we're there. It is where Anne hides
> to survive.
>
> —JAMES MOLLOY, *Bard College, class of 2010*

LATE IN THE FALL OF 2007, I TAUGHT *The Diary of Anne
Frank* to a class at Bard College. It was a course in close reading,
in which we'd been studying the works of writers ranging from
John Cheever to Hans Christian Andersen, from Mavis Gallant
to Leonard Michaels, from Roberto Bolaño to Grace Paley. My
students were not only intelligent, passionate, and engaged, but
intuitive and remarkably well read, and I was often surprised
and delighted by the leaps of imagination and association that
led them from literature to the visual arts or music. A discus-

sion of Bolaño had turned into a conversation about Borges. A class on Andersen's "The Snow Queen" had inspired a discussion of the innocent, perverse, fairy-tale eroticism of the self-taught artist Henry Darger.

I was eager to hear what they would say about Anne Frank, but I wasn't—nor were they—prepared for the intensity of their responses. I had been thinking and writing about the benefits and the risks of identifying with Anne Frank; my students demonstrated all of the former and none of the latter. Born long after her death, they felt as if she were speaking to them. As if she were one of them. They identified with her humanity, her sympathy, her humor, her impatience, her alienation, her adolescent struggles, without ever losing sight of the gap between their comfortable and privileged lives and the circumstances that had driven her into hiding. They were keenly aware of the gap between what Anne was forced to endure and the trivial setbacks that their contemporaries found nearly unendurable.

A student wrote, "I couldn't believe how she kept resolving to be happier. She writes about a lot of experiences of joy. Even in those extremes she manages to maintain the psyche of a normal girl. Today such little things turn people into basket cases, they go on Prozac because they can't pay their credit card bills. It's hard to believe that she manages to maintain so much of herself. She can look out at Amsterdam on a sunny day and still be transfixed by beauty."

I had asked them to send me brief response papers in advance of the class, and, perhaps because we'd placed so much emphasis on *how* writers wrote, quite a few of them focused on Anne Frank's eloquence.

Not only was she a fabulous writer, but I felt a special connection to her because my grandparents were in hiding

during the war, in France. At one point she says she wants to
be a journalist, and I kept thinking that this is one of the best
journalistic documents in history. She knows so much. She
noticed all the warning signs, Jews are not allowed to do this,
Jews are not allowed to do that. When her sister was called
up, everyone knew what that meant. It's amazingly beauti-
fully written, and she does such a good job of making you feel
the fear that was at the base of everything, all the time.

Wrote another student, "She creates characters so believ-
able I had to keep reminding myself that they were real." Another noted, "There's something eerie and amazing about the
level and the kind of details she gives us. Dialogue in chunks,
descriptions of actions, and everywhere, character character
character. This girl is an amazing writer. I find myself wondering, did she know what she was doing? It's clear that Anne
wrote this diary for herself, and it meant a lot to her, but was it
ever anything else? Isn't all writing inevitably 'something else,'
meaning, it's not just for the writer? How can the act of writing
not be for someone else? Is it possible to write, to tell a story,
without thinking of someone you're telling it to?"

In class, I encouraged them to talk about the difference between their first encounter with the diary—most of them had
read it on their own, or had been assigned to read it in junior
high or high school—and how it seemed to them now, especially after having taken a class in close reading. One young
woman wryly remarked that she hadn't read the diary before
because she'd grown up in Seattle, where "we did the Japanese
internment camps instead." A few admitted, with embarrass-
ment, that, even though they'd read the diary in high school,
they'd had no idea how Anne died, and were horrified to have
finally learned the truth.

Nearly all of those who had read it before mentioned that

they'd previously had almost no sense of the book as literature but only as a historical document, or as some sort of young-adult coming-of-age memoir. One student said he'd gotten funny looks and sarcastic remarks from his Bard schoolmates (not those enrolled in the class) when they saw him reading Anne Frank's diary. They acted as if he were assuming some sort of ironic-regressive pose that involved carrying around a children's classic, the equivalent of using his grade school lunch box as an attaché case.

Nearly all of them talked about how (perhaps because they'd been so young themselves) they'd had little sense of Anne as a character—and, specifically, of how much she had changed and grown in two years. "What got to me," said one young man, "is that she starts out as a little kid and matures and can see things more objectively. Instead of being mad at people she can step back and see herself. She comes to be this really wonderful human being. I loved the way it ends with her thinking how good life would be without other people in it. It makes it more tragic that she couldn't fulfill all the talent and humanity that she had."

Another agreed. "She started out as such an innocent optimistic girl and she became so much more self-conscious and self-aware."

Yet another wrote, "Anne is stunning. She is so powerfully alive. (To phrase it this way sounds a little stupid to me but I'm not sure how else to say it.) Everything she describes about their Secret Annex is interesting because *she* is interacting with it, and telling about it in her insightful, hilarious way. She often talks about her parents, Mr. and Mrs. Van Daan and Dussel trying to discipline her or shame her into changing her behavior. We never actually see what happens before the adults go after her, because Anne is always writing after the fact, not accounting for her own behavior, whatever she did, but it is not hard to imagine. Anne was probably very hard to be around! She talked constantly; she spoke her mind, she felt strongly, and was only a thirteen-year-old. What a person to live with in a Secret Annex!

"We get to be 'Kitty,' the friend she invents and addresses all of her entries to. Anne meets us for the first time and she slowly gets to know us and feel comfortable confiding in us. As Kitty, we are the depository for her secrets."

Except for one young woman, who had somehow gotten hold of the entire *Critical Edition* and realized (as most readers have not) the implications of the "a," "b," and "c" versions, my students were amazed to hear that Anne had gone back and revised her journal. But as soon as they'd had a chance to think about that, they felt, as I did, that this made the diary more impressive rather than less authentic.

As the December dusk deepened outside the window and the classroom (I tried to avoid switching on the harsh fluorescent lights) grew darker, the students' voices grew quieter, and they seemed sobered and saddened as they spoke about Anne's last days. I kept thinking that these were precisely the sort of open-hearted, idealistic young people who might someday wind up working at a place like the Anne Frank Foundation, trying to improve our damaged and possibly doomed world. The tone in which they spoke about the diary evoked that of a eulogy, or of testimony. It *was* as if they were talking about a friend. One of my students summarized the essence of what he'd written in his response paper:

> *I wonder if it is just me or if her writing is so personal that it is symptomatic of every reader, but I feel an emotional connection to Anne Frank through her writing. This is a testament to both the power of her writing and her character. To feel a real connection with a girl who has been dead for*

almost 63 years . . . is a strange emotional experience, but I feel as though I know her well. I know that, given the chance, we would have been close in life. We have a lot in common in terms of our interests and desires. She and I both love writing and history and loathe math and figures, and I admire her deep sense of self-awareness and her emotional transparency that is evident in her writing. She has a passion for self-expression I find very moving, and I wish I could be that honest and clear when I write for myself . . . Anne and I also share a passion for nature and recognize the power in the simple beauty of everyday experiences of nature. I can picture many a night where I have stared out my window in the same way; the beauty of the night filling me with excitement and keeping me awake . . . I think I may have read this book in grade school but I am very glad to have a fresh look at it again now that I am 20 years old. There is so much here that passed me by.

Listening to him, I thought: They *would* have been friends. She was a fifteen-year-old girl. She saw herself as both ordinary and special, growing up under circumstances that were in no way normal even as her parents insisted on going through the motions of everyday life. What was certain was that Anne did not grow up believing that she was going to be sent to Auschwitz, and die, at fifteen, in Bergen-Belsen.

I listened to my students, as fresh and eager as she had been, only a few years older than she was when she died. I asked one of them to read aloud from the diary, and he chose the final entry, the passage in which Anne imagined the person she could have been if there weren't any other people living in the world.

When he finished, the class was silent. In the hush, I thought about Anne's wish to go on living after her death. And it was clear to me, as it has been throughout the writing of this book, that her wish has been granted. I remembered how, more than fifty years ago, the first time I read the diary, I'd kept reading until the light had faded in my bedroom, as it had now, in this classroom. And for those few hours during which my students and I talked about her diary, it seemed to me that her spirit—or, in any case, her voice—had been there with us, fully present and utterly alive, audible in yet another slowly darkening room.

ACKNOWLEDGMENTS

AMONG THE THINGS I LEARNED IN THE PROCESS OF
writing this book is what writers mean when they say, on
pages such as this one, that there are people without whom
their work would simply not have been possible. Below is a
list (only partial, I fear) of those who made it a possibility—
and a pleasure—for me to conceptualize and complete this
book.

In Amsterdam, Annemarie Bekker, Mariela Chyrikins,
Theresien Da Silva, and Norbert Hinterleitner gave generously
of their time. Erika Prins not only guided me through the intricacies of the Anne Frank archive, but read the manuscript
and made helpful suggestions. Jan Erik Dubbelman and Dienke
Honduis were unfailing sources of friendship, expertise, and
encouragement when it was most needed. A historian and a
fellow writer, Dienke gave my manuscript a meticulous reading and offered useful and indeed essential criticisms and edits
that I incorporated in the final draft.

In Basel, Bernd Elias and Barbara Eldridge read the book
with alacrity and responded with an enthuasism that meant

a great deal to me. At the Anne Frank Center in New York, Maureen McNeil not only introduced me to the teachers and students at the Bell Academy, but provided invaluable introductions to her colleagues in Amsterdam.

I would like to thank Peter Carey and the Hertog Fellowship program of the MFA program at Hunter College, which provided research assistants who aided me at every stage: Ana Jomolca, Annie Levin, Tennessee Jones. Thanks also to Zachary Wolfe, Christina Bailly, and Alexandra Bowe for helping to ready the book for publication, and to Mark Schaevers for his humor, his friendship, and his assistance with research and translation.

My brilliant students at Bard College reacted to Anne Frank's diary in ways that moved me so deeply that I decided to end the book with a chapter describing their responses. I would like to thank each and every one of them: Alex Carlin, Gabriel DeRita, Evelyn Fettes, Sam Freilich, Simon Glenn Gregg, Shay Howell, Samuel Israel, Sonya Landau, Sara Lynch-Thomason, James Molloy, Emily Moore, Evan Neuwirth, Angela Sakrison, Tegan Walsh, and Daniel Whitener. I would also like to thank Leon Botstein for bringing me to Bard, and Norman Manea for his friendship and his kindness in introducing me to this remarkable institution.

My editor, Terry Karten, was, as always, patient, inspiring, and more helpful (in more ways) than I can possibly say. Nor can I express how profoundly I depended on the resourcefulness, the thoughtfulness, and the unfailingly cheerful and positive energy of my agent, Denise Shannon. Though I used to joke that I could hear my friends picking up a magazine on the other end of the phone when I began to rant obsessively about the subject matter of this book, the truth is that I relied on them to listen and advise me. Thanks to my sons, Bruno and Leon Michels, and my daughter-in-law, Yesenia Ruiz. And finally, not one word of this book, or of anything I have written for more than thirty years, would have ever made it onto the page without the love and advice and support of my husband, Howie Michels.

SOME NOTES ON THE TEXT

ONE OF THE COMPLICATIONS IN WRITING ABOUT ANNE Frank's diary is created by the fact that she gave pseudonyms to the people with whom she shared the secret annex. Throughout the text, I have used the real names of the historical figures—for example, the Van Pels family—in place of the names these figures have been given in Anne's diary—for example, the Van Daans. Exceptions occur when I am quoting from the *Diary*, or when I am referring to the characters in the play and the film, in which the real-life models are known exclusively by their pseudonyms.

In the Netherlands, the Anne Frank House (that is, the building in which the secret annex is located) is known as the Anne Frank Museum. The organization that supports the museum and the human-rights programs associated with Anne Frank and her diary is called, in Dutch, the Anne Frank Stichting. Though *Foundation* is not a direct translation of *Stichting*, I have used the term *Foundation* for clarity. Also, as I have made clear, the Anne Frank-Fonds in Basel is a separate organization.

NOTES

The following abbreviation is used in the notes:

DAF: *The Diary of Anne Frank: The Revised Critical Edition*. Prepared by the Netherlands State Institute for War Documentation. Edited by David Barnouw and Gerrold van der Stroom, translated by Arnold J. Pomerans and B. M. Mooyaart-Doubleday (New York: Doubleday, 2003).

CHAPTER 1 *The Book, The Life, The Afterlife*

6   **"Anne would . . ."** Interview with Hanneli Pick-Goslar on Scholastic Web site, www.scholastic.com.

6   **"If I haven't any talent . . ."** *DAF,* April 4, 1944.

7   **"I saw that Anne was writing . . ."** Miep Gies, *Anne Frank Remembered*, with Alison Leslie Gold (New York: Simon and Schuster, 1987), 186.

7   **"whether Anne Frank has *had* any serious readers . . ."** John Berryman, *The Freedom of the Poet* (New York: Farrar, Straus & Giroux) 92.

7   **"One thing is certain . . ."** G. B. Stern, "Introduction to *Tales from the House Behind,* Kingswood, England, 1952. Harry Mulisch, "Death and the Maiden," in *Anne Frank: Reflections on her Life and Legacy*, eds. Hyman A. Enzer and Sandra Solatoroff-Enzer (Urbana and Chicago: University of Illinois Press, 2000), 81.

8   **"A child's diary . . ."** Harold Bloom, *A Scholarly Look at the Diary of Anne Frank* (New York: Chelsea House, 1999), 1.

8   **"The work by this child . . ."** Reprinted in Enzer, *Anne Frank: Reflections on Her Life and Legacy*, 96.

8   **"I do not mean . . ."** Robert Alter, "The View from the Attic: An Obsession with Anne Frank," the *New Republic,* December 4, 1995, 58.

11   **"Of course . . ."** *DAF,* March 29, 1944.

11   **"kept her company . . ."** Philip Roth, *The Ghost Writer* (New York: Farrar, Straus & Giroux, 1979), 136.

12   **"Just imagine . . ."** *DAF,* March 29, 1944.

12   **"I must work . . ."** *DAF,* April 4, 1944.

12   **"Everything here . . ."** *DAF,* April 14, 1944.

13   **"Whether these leanings . . ."** *DAF,* May 11, 1944.

13   **"At long last . . ."** *DAF,* May 20, 1944.

14   *"The Diary of a Young Girl is not . . ."* Mirjam Pressler, *Anne Frank: A Hidden Life* (New York: Puffin Books, 2000), 15.

14   **"That ingenuous title . . ."** Judith Thurman, "Not Even a Nice Girl," *Cleopatra's Nose: 39 Varieties of Desire* (New York: Farrar, Straus & Giroux, 2007), 101.

14   **"if we take . . ."** H. J. J. Hardy, "Documents Examination and Handwriting Identification of the Text Known as the *Diary of Anne Frank*: Summary of Findings," *DAF, The Revised Critical Edition,* 166.

15   **"I am the best . . ."** *DAF,* April 4, 1944.

17   **"What a comparison . . ."** Thurman, "Not Even a Nice Girl," 99.

17   **"Otherwise, it would have seemed . . ."** Interview with David Barnouw, December 2007.

18   **She was the first to note** . . . Laureen Nussbaum, "Anne Frank," in Enzer, *Anne Frank: Reflections,* 21–31.

20   **"Five Precious Pages . . ."** Ralph Blumenthal, "Five Precious Pages Renew Wrangling Over Anne Frank," *New York Times,* September 10, 1998, A6.

CHAPTER 2 *The Life*

23   **"I don't want to set down . . ."** *DAF,* June 20, 1942.

24   **"Miep had ten drinks . . ."** *DAF,* May 8, 1944.

25   **"Miep made our mouths . . ."** *DAF,* May 8, 1944.

26   *"as we are Jewish . . ."* *DAF,* June 20, 1942.

28   **"God knows . . ."** Melissa Müller, *Anne Frank, The Biography,* trans. Rita and Robert Kimber (New York: Henry Holt, 1988), 132.

29   **"From then on . . ."** Interview with Hanneli Pick-Goslar on Scholastic Web site, www. scholastic.com.

29   **"always fussing"** Willi Lindwer, *The Last Seven Months of Anne Frank* (New York: Random House, 1991), 16.

30   **"Once, when Mutti . . ."** Eva Schloss with Evelyn Julia Kent, *Eva's Story* (Great Britain: W. H. Allen and Co., 1988), 32.

30   **"When Anne . . ."** Ernst Schnabel, *Anne Frank, A Portrait in Courage* (New York: Harcourt, 1958), 49.

32 "**knew her only from Anne's girlhood days . . .**" Schnabel, 58.

32 "**The rest of our family**" . . .   *DAF,* June 20, 1942.

33 "**All the correspondents . . .**" Geert Mak, *Amsterdam,* trans. Philip Blom (London: The Harvill Press, 1999), 249.

35 "**Our North Sea . . .**" Dick Van Galen Last and Rolf Wolfswinkel, *Anne Frank and After: Dutch Holocaust Literature in a Historical Perspective* (Amsterdam: Amsterdam University Press, 1996), 43.

35 "**three marvelous days**" Gies, *Anne Frank Remembered,* 68.

36 "**The school . . .**" Anne Frank, *Tales from the Secret Annex,* trans. Susan Masotty (New York: Bantam Books, 2003), 52.

38 **One teacher** . . . Dr. J. Presser, *The Destruction of the Dutch Jews* (New York: E.P. Dutton & Co., 1969), 127.

39 "**So we could not do . . .**" *DAF,* June 20, 1942.

40 "**Under proper guidance . . .**" Heydrich, Minutes of Wannsee Conference.

41 "**At the end . . .**" Last and Wolfswinkel, *Anne Frank and After: Dutch Holocaust Literature in a Historical Perspective,* 54.

42 "**Often one made . . .**" Last and Wolfswinkel, 45.

43 "**Everybody had a family . . .**" Last and Wolfswinkel, 69.

43 "**were a pleasure . . .**" Last and Wolfswinkel, 10.

44 "**very blond young woman**" Gies, *Anne Frank Remembered,* 44.

45 "**Jews were such an established part . . .**" Gies, 29.

45 "**There is a look . . .**" Gies, 88.

47 "**We would be going . . .**" *DAF,* July 9, 1942.

48 "**A few days later . . .**" Schnabel, *Anne Frank, A Portrait in Courage,* 103.

49 "**Primarily, however . . .**" www. Annefrank.org.

50 "**The whole day . . .**" *DAF,* July 10, 1942.

51 "**In the evenings . . .**" *DAF,* November 19, 1942.

52 "**Another thing . . .**" *DAF,* Sept 16, 1943.

52 "**concrete results**" Harry Paape, "The Betrayal," in *Revised Critical Edition.* p. 40.

54 "**Will the boy . . .**" Etty Hillesum, *An Interrupted Life* (New York: Henry Holt, 1996), 208.

54 "**That was very messy . . .**" Lindwer, *The Last Seven Months of Anne Frank,* 1988, 52.

56 "**'Anne, who was already sick. . .**" Schnabel afterword to *Diary of a Young Girl,* 281.

57 "**had little squabbles . . .**" Lindwer, *The Last Seven Months of Anne Frank,* 104.

58 "**Anne stood in front of me . . .**" Lindwer, 74.

CHAPTER 3 *The Book, Part I*

65 "**as though he might . . .**" Schnabel, *Anne Frank, A Portrait in Courage,* 133.

67   "... zero was Anne Frank." Simon Wiesenthal, *The Murderers Among Us* (New York: McGraw-Hill, 1967), 171–183.

70   "I am not ..." Gies, *Anne Frank Remembered*, 11.

70   "Never have we heard ..." *DAF,* January 28, 1944.

70   "You can see ..." *DAF,* May 8, 1944.

73   "He stood on the porch and rang ..." Jon Blair, Director, *Anne Frank Remembered*, Sony Pictures, 1995.

74   "I could tell ..." Gies, *Anne Frank Remembered*, 235.

76   "If we bear ..." *DAF,* April 11, 1944.

77   "the most moving ..." Gerrold van der Stroom, "The Diaries, *Het Achterhuis* and the Translations," *DAF,* 64.

78   "witless barbarity" ... "When I had finished ..." Van der Stroom, *DAF,* 67.

79   "diary of a normal child ..." Romein, introduction, *Het Achterhuis* (Amsterdam: Contact, 1947), trans. Mark Schaevers.

80   "A book intended ..." Van der Stroom, *DAF,* 73.

80   "It is an interesting document ..." Letter to Otto Frank in Anne Frank archive.

82   "One day ..." Judith Jones, *The Tenth Muse, My Life in Food* (New York: Knopf, 2007), 46.

82   "Let me say again ..." Letter to Otto Frank in Anne Frank archive.

83   "I love the book ..." Letter to Otto Frank in Anne Frank archive.

83   "Jewish Joan of Arc" Ian Buruma, "The Afterlife of Anne Frank," the *New York Review of Books,* February 19, 1998, 4.

85   "What she has left behind ..." Antonia White, review of *The Diary of a Young Girl,* by Anne Frank, *The New Statesman,* May 1953.

85   "very charming" Letter from Donald Elder to Otto Frank in Anne Frank archive.

86   "This is a remarkable ..." *DAF,* introduction.

86   "... jocular ... *away* from the Jews." Geoffrey Ward, *A First-Class Temperament* (New York: HarperCollins, 1992), 251, 661.

86   "Reading your introduction ..." Letter from Otto Frank to Eleanor Roosevelt in Anne Frank archive.

87   "mission in publishing ..." Letter from OF to ER in Anne Frank archive.

87   "For little Anne Frank. ..." Meyer Levin, "The Child Behind the Secret Door," the *New York Times Book Review,* June 15, 1952, 1.

88   "one of the most moving ..." *Time* magazine, June 16, 1952.

88   "extraordinary ..." *Commonweal* Volume 6, no. 12, June, 1952.

88   "ANNE ..." Letter from Barbara Zimmerman to Otto Frank in Anne Frank archive.

CHAPTER 4 *The Book, Part II*

90   Pressler, *Anne Frank: A Hidden Life,* 31.

92   "I twist ..." *DAF,* August 1, 1944.

94   "I have one outstanding . . ." *DAF*, July 15, 1944.
94   "Then a certain person . . ." *DAF*, August 4, 1943.
95   "I see the eight of us . . ." *DAF*, November 8, 1943.
97   "Harry visited us yesterday . . ." *DAF*, July 3, 1942.
97   "We ping-pongers . . ." *DAF*, June 20, 1942.
98   "She was vivacious . . ." Berryman, *The Freedom of the Poet*, 95.
100 "Although it is fairly warm . . ." *DAF*, May 18, 1943.
100 "If the conversation . . ." *DAF*, January 28, 1944.
101 "daily timetable" *DAF*, August 9, 1943.
105 "Following Daddy's good example . . ." *DAF*, October 29, 1942.
106 "She talks so unfeelingly . . ." Masotty, *Tales from the Secret Annex*, 162.
106 "It isn't sentimental nonsense . . ." *DAF*, May 11, 1944.
107 "When he complained . . ." *DAF*, March 10, 1943.
107 "Just as I shrink . . ." Van der Stroom, *DAF*, p. 77.
118 "Presumably man . . ." *DAF*, June 15, 1944.
109 "I was very unhappy again . . ." *DAF*, December 29, 1943.
110 "If I think . . ." *DAF*, March 7, 1944.
111 "She quarrels . . ." *DAF*, June 16, 1944.
111 "a very uncomplicated person . . . If anyone . . ." Schnabel, *Anne Frank, A Portrait in Courage*, 106.
111 "way to express . . ." Gies, *Anne Frank Remembered*, 173.
112 "The yells and screams . . ." *DAF*, October 29, 1943.
112 "Mrs. Van Daan, the fatalist . . ." *DAF*, March 10, 1943.
114 "It gave me a queer feeling . . ." *DAF*, January 6, 1944.
117 "Otto and Edith's decision . . ." Müller, *Anne Frank, The Biography*, 188.
118 "sharing out . . ." *DAF*, May 6, 1944.
118 "When he has ended . . ." *DAF*, December 22, 1942.
118 "Not only did his hair . . ." *DAF*, December 22, 1943.
119 "most hoity-toity . . ." *DAF*, February 14, 1944.
119 "Dussel thinks . . ." *DAF*, July 23, 1943.
119 "I think it's so rotten . . ." *DAF*, March 20, 1944.
120 "I don't want to be in the least . . ." *DAF*, February 5, 1943.
120 "The only things that go down . . ." *DAF*, August 9, 1943.
120 "I would want to have the feeling . . ." *DAF*, March 20, 1944.
121 "Just because . . ." *DAF*, November 7, 1942.
122 "I have seldom . . ." Berryman, *The Freedom of the Poet*, 95.
123 "like a little island . . ." *DAF*, November 9, 1942.
123 "Mrs. Van Daan sat bolt upright . . ." *DAF*, May 18, 1943.
123 "The room was in a glorious . . ." *DAF*, December 10, 1942.
125 "I wasn't quite my usual . . ." *DAF*, January 24, 1944.
127 ". . . Rauter . . ." *DAF*, March 27, 1943.
127 "Oh, Kitty . . ." *DAF*, June 6, 1944.

CHAPTER 5 *The Book, Part III*

130 "From Mummy and Daddy . . ." *DAF,* June 14, 1942.

131 "Miss J. always has . . ." *DAF,* June 15, 1942.

131 "We went to oasis . . ." *DAF,* June 30, 1942.

134 " 'But, Daddy . . ." *DAF,* July 5, 1942.

135 "Anne, putting herself . . ." Nussbaum, "Anne Frank," in Enzer, *Anne Frank: Reflections,* 26.

137 "I couldn't refrain. . . ." *DAF,* January 6, 1944.

137 "From early in the morning . . ." *DAF,* February 27, 1944.

138 "Peter has touched . . ." *DAF,* April 28, 1944.

139 "When revising . . ." Nussbaum, "Anne Frank," in Enzer, *Anne Frank: Reflections,* 28.

140 "how the houses . . ." *DAF,* March 29, 1944.

142 "One Sunday morning . . ." *DAF,* September 28, 1942.

146 "In this first book . . ." Sylvia P. Iskander, "Anne Frank's Reading: A Retrospective," in Enzer, *Anne Frank: Reflections,* 103.

147 "After he'd been working . . ." *DAF,* October 20, 1942.

148 "Mr. van Pels repeated . . ." *DAF,* August 14, 1942.

149 "Prospectus and Guide . . ." *DAF,* November 17, 1942.

150 "on November 27, 1943, Anne writes . . ." Pressler, *Anne Frank: A Hidden Life,* 140.

153 "The little hole underneath is so terribly small . . ." *DAF,* March 24, 1944.

154 "Otto Frank had picked . . ." Nussbaum, "Anne Frank," in Enzer, *Anne Frank: Reflections,* 24.

154 "A reader . . ." Nussbaum, 30–31.

155 "When I look over . . ." *DAF,* January 22, 1944.

CHAPTER 6 *The House*

164 "The moment . . ." Interview with author Mariela Chyrikins, Amsterdam, February 14, 2008.

166 "If you're surprised . . ." Interview with Norbert Hinterleitner, Amsterdam, February 14, 2008.

166 "True, the ending happens . . ." Bruno Bettleheim, "The Ignored Lesson of Anne Frank," in Enzer, *Anne Frank: Reflections,* 189.

167 "kitsch . . ." Ian Buruma, "The Afterlife of Anne Frank," the *New York Review of Books,* February 19, 1998, 4.

168 "The line . . ." Lawrence Langer, "The Americanization of the Holocaust on Stage and Screen," in Enzer, *Anne Frank: Reflections,* 201–2.

169 "Anyone who claims . . ." *DAF,* July 15, 1944.

170 "been torn . . ." Cynthia Ozick, "Who Owns Anne Frank?," *The New Yorker,* October 6, 1997, 78.

170 "not gruesome enough" Sem Dresden, *Persecution, Extermination,*

Literature, trans. Henry G. Shogt (Toronto: University of Texas Press, 1995), 198.

170 **"The diary is taken to be . . ."** Ozick, "Who Owns Anne Frank?," 78.

170 **"A girl's journal . . ."** Alter, "The View from the Attic: An Obsession with Anne Frank," the *New Republic,* December 4, 1995, 58.

171 **"Statistics don't bleed . . ."** Arthur Koestler, *The Yogi and the Commissar and other essays* (London: Hutchinson & co., 1985), 97.

171 **"One single Anne Frank . . ."** Primo Levi, *The Drowned and the Saved* (New York: Random House, 1988), 145.

171 **"If Anne Frank's diary . . ."** Ludwig Lewisohn, review of *The Diary of A Young Girl,* by Anne Frank, the *Saturday Review,* July 1952, p. 20.

172 **"Such identification . . ."** Buruma, "The Afterlife of Anne Frank," *NYRB,* 4.

172 **"Despite . . ."** Ozick, "Who Owns Anne Frank?", 79.

173 **"The unabashed . . ."** Ozick, 80.

174 **"deeply truth-telling . . ."** Ozick, 78.

CHAPTER 7 *The Play*

179 **"from amongst . . ."** Meyer Levin, *The Obsession* (New York: Simon and Schuster, 1974), 28.

181 **"he made just one . . ."** Judith Jones, *The Tenth Muse,* 46.

181 **"In the middle of life . . ."** Levin, *The Obsession,* 7.

182 **"Levin has the hallucination . . ."** Levin, 37.

182 **"Agreed, it was an obsession . . ."** Levin, 13.

182 **"ringed by eternal fire . . ."** Levin, 39.

182 **"You have been my Hitler"** Levin, 40.

183 **"Again and again . . ."** Levin, 31.

183 **"Anne Frank's diary . . ."** Meyer Levin, *The New York Times Book Review,* "Life in the Secret Annex," June 15, 1952.

185 **"agenting the tome . . ."** *Variety,* June 18, 1952.

187 **"screwing up . . ."** Letter from Barbara Zimmerman to Frank Price, quoted in Ralph Melnick, *The Stolen Legacy of Anne Frank: Meyer Levin, Lillian Hellman, and the Staging of the Diary* (New Haven: Yale University Press, 1997), 22.

187 **"impossible . . . play,"** Barbara Zimmerman letter, quoted in Lawrence Graver, *An Obsession with Anne Frank: Meyer Levin and the Diary* (Berkeley: University of California Press, 1997), 72.

189 **"If he writes the play . . ."** Letter from Meyer Levin to Otto Frank, quoted in Melnick, *The Stolen Legacy,* 24–5.

189 **"I always said . . ."** Letter from OF to ML, quoted in Graver, *An Obsession with Anne Frank,* 54.

192 **"castrating homosexual"** Letter from ML to OF, quoted in Graver, 50.

193 **"The only way . . ."** David L. Goodrich, *The Real Nick and Nora: Fran-*

ces Goodrich and Albert Hackett, Writers of Stage and Screen Classics (Carbondale: Southern Illinois University Press, 2004), 207.

194 **"big-name dramatist . . ."** Levin, *The Obsession,* 61.

194 **"The very origin . . ."** Levin, 36.

195 **"Quite a triumph . . ."** Meyer Levin, "Anne Frank: A Play," unpublished ms. in the Dorot Jewish Division, the New York Public Library, 2.

195 **"We must go into hiding . . ."** Levin, "Anne Frank: A Play," 7.

196 **"Never let a man choose a house"** Levin, 17.

196 **("It's part of being something . . ."** Levin, 68–69.

197 **"I suppose it could be . . ."** Levin, 41.

198 **"When you were at home . . . that's God"** Anne Frank, *Tales from the Secret Annex,* 172–3.

198 **"End of the diary"** Levin, "Anne Frank: A Play," 60.

199 **"I think I have never . . ."** Letter from Carson McCullers to Otto Frank, quoted in Graver, *An Obsession with Anne Frank,* 51–2.

200 **"We have no . . ."** Letter from CM to OF in Anne Frank archive.

200 **"In spite of our . . ."** Letter from CM to OF in Anne Frank archive.

200 **"by the Nazis . . ."** Letter from ML to OF, quoted in Melnick, *The Stolen Legacy,* 88.

201 **"moments of lovely comedy . . ."** Goodrich and Hackett papers, quoted in Graver, *An Obsession with Anne Frank,* 78.

201 **"tremendous responsibility"** Hackett, "Diary of the Diary of Anne Frank," *New York Times,* September 30, 1956, xi.

202 **"A Challenge . . ."** *New York Post,* January 13, 1954, quoted in Graver, *An Obsession with Anne Frank,* 80.

204 **"cavalier"** Letter from ML to OF, quoted in Melnick, *The Stolen Legacy,* 104.

205 **"brilliant"** Hackett, "Diary of the Diary," xi.

205 **"embarrassing . . ."** Garson Kanin to Frances and Albert Hackett, quoted in Melnick, *The Stolen Legacy,* 115.

206 **"I thought I could not cry . . ."** Hackett, "Diary of the Diary," Arts and Leisure, page xi.

207 **"In all my meetings . . . crashed down."** Bernard Kalb, "Diary Footnotes," *New York Times,* October 2, 1955, 3.

207 **Warsaw ghetto . . .** Letter from ML to OF, quoted in Melnick, *The Stolen Legacy,* **123.**

207 **"flamboyant and dashing"** Joseph Schildkraut, *My Father and I* (New York: Viking Press, 1959), 230.

208 **"uncanny"** Hackett, "Diary of the Diary," xi.

208 **"had first-hand . . ."** *The Diary of Anne Frank,* directed by George Stevens, 20th Century Fox, 1959, DVD, supplemental material.

208 **"This is not a play . . ."** Hackett, "Diary of the Diary," xi.

209 **"I told him . . ."** Bernard Kalb, "Diary Footnotes," *New York Times,* October 2, 1955, xi.

209 **"Both Kermit . . ."** Hackett, "Diary of the Diary," xi.

209 **"Mostly, though . . ."** Kalb, "Diary Footnotes," xi.

213 **"You are the most intolerable . . ."** Goodrich and Hackett, *The Diary of Anne Frank and Related Readings* (Evanston: McDougall Littell, 1997), 30.

215 **"Some day . . ."** Goodrich and Hackett, 118.

216 **"going through a phase . . ."** Goodrich and Hackett, 118.

216 **"She puts me to shame."** Goodrich and Hackett, 122.

216 **"they have made . . ."** Brooks Atkinson, review of *The Diary of Anne Frank, New York Times,* October 6, 1955, 24.

217 **"There is only one way . . ."** Brooks Atkinson, "Inspired Theater," *New York Times,* October 16, 1955.

217 **"the punch of plain . . ."** *Newsweek,* October 17, 1955, 103.

217 **"I can think . . ."** The *New Yorker,* October 15, 1955, 75–6.

218 **"The house lights . . ."** Kenneth Tynan, "At the Theater: Berlin Postcript," *London Observer,* November 7, 1956.

218 **"Yes, but *that* girl . . ."** Adorno, Theodor W., "What Does Coming to Terms with the Past Mean?" in *Bitburg in Moral and Political Perspective,* ed. Geoffrey H. Hartman (Bloomington: Indiana University Press, 1986) 127.

219 **"I wanted to restore . . ."** Interview with Wendy Kesselman, telephone, summer, 2007.

221 **"To see Natalie . . ."** Ben Brantley, "This Time, Another Anne Confronts Life in the Attic," *New York Times,* December 5, 1997, 16.

221 **"Despite the changes, . . ."** Molly Magid Hoagland, "Anne Frank Onstage and Off," *Commentary,* March 1998.

221 **"earnestly artificial . . ."** Vincent Canby, "A New Anne Frank Still Stuck in the 50's," *New York Times,* December 21,1997, 5.

223 **"At 16 . . ."** Natalie Portman, "Thoughts from a Young Actor," *Time* magazine, June 14, 1999, 80.

CHAPTER 8 *The Film*

226 **"I was seventeen . . ."** Anne Frank, *Tales from the Secret Annex,* 82.

229 **"How many little girls . . ."** *The Diary of Anne Frank,* directed by George Stevens, 20th Century Fox, 1959, DVD, supplemental material.

231 **"He approved of me . . ."** *The Diary of Anne Frank,* directed by George Stevens.

231 **"no greater . . ."** *The Diary of Anne Frank,* directed by George Stevens.

232 **"You have given . . ."** Judith Doneson, *The Holocaust in American Film* (New York: Syracuse University Press, 2002), 72.

234 **"If it's that bad . . ."** *DAF,* October 9, 1942.

234 **"I didn't understand . . ."** *The Diary of Anne Frank,* directed by George Stevens.

235 **"accessible . . ."** *The Diary of Anne Frank,* directed by George Stevens.

235 "I watched . . ." Shelley Winters, *Shelley II, The Middle of My Century* (New York: Simon and Schuster, 1998), 227.

236 "And so it seems . . ." *The Diary of Anne Frank*, directed by George Stevens.

237 "But Anne . . ." *The Diary of Anne Frank*, directed by George Stevens.

Chapter 9 Denial

239 "When a . . ." Robert Boatman, letter to the *Idaho Statesman*, July 1998.

243 "So you are . . ." Teressa Hendry, the *American Mercury*, summer 1967.

246 "the first pedophile . . ." Ditlieb Felderer, *Anne Frank's Diary, A Hoax* (Torrance, CA: Institute for Historical Review, 1979), 64.

246 "Apparently . . ." Felderer, 6.

248 "We never had illusions . . ." Francis X. Clines, "Anne Frank Again Focus of Challenge," *New York Times*, April 21, 1987, section A, page 10.

249 "symbolically free himself . . ." German Press Agency, February 26, 2007.

Chapter 10 Teaching the Diary

253 "The *Diary* is many things . . ." Dorothea Shefer-Vanson, *The Diary of Anne Frank*, Cliffs Notes (Hoboken, NJ: Wiley Publishing, 1984), 6.

256 "Perhaps teachers hesitate . . ." Rebecca Kelch Johnson, "Teaching the Holocaust," in the *English Journal*, vol. 69, no. 7, October 1980, 69.

256 "There's in people . . ." *DAF*, May 5, 1944.

256–57 "champions goodness . . ." Lesley Shore, "Anne Frank in Life and Death: Teaching the Lesson of the Holocaust," from Proceedings of the 38th Annual Convention of Jewish Libraries, Toronto, June 15–28.

257 True or false . . . Michelle Keller, "Remembering the Holocaust," Education Resources Information Center, teacher classroom guide, 2002.

258 "Hitler's 'final . . ." Sue Jones Erlenbusch, *Projects for Anne Frank: Diary of a Young Girl*, 1993, www.teachervision.com.

258 1. I want my memory . . . Mari Lu Robbins, *A Guide for Using Anne Frank* The Diary of a Young Girl *in the Classroom* (Westminster, CA: Teacher Created Resources, 2007), 12.

259 "Even when students . . ." Karen Spector and Stephanie Jones, "Constructing Anne Frank: Critical Literacy and the Holocaust in Eighth-Grade English," in *Journal of Adolescent and Adult Literacy*, September 2007, 40.

260 "ask them to reflect . . ." Karyn M. Peterson, Cyberhunt Teacher's Page, May–June 2004, www.scholastic.com.

261 **"what literature is . . ."** Robert Probst, "Literature as Invitation," in *Voices from the Middle,* vol. 8, no. 2, December 2000, 8–15.

263 **("Let us take . . ."** Judith Tydor Baumel, "Teaching the Holocaust through the *Diary of Anne Frank,"* in *Anne Frank in Historical Perspective: A Teaching Guide for Secondary Schools,* eds. Alex Grobman and Joel Fishman (Los Angeles: Martyrs Memorial and Museum of the Holocaust of the Jewish Federation Council of Greater Los Angeles, 1995), 49.

265 **"At Fowler High School . . ."** National Coalition Against Censorship, www.ncac.org.

267 **"It is this underlying . . ."** *Bob Mozert et al. v. Hawkins County Public Schools et al.* U.S. District Court for the Eastern District of Tennessee, Northeastern Division, 1984.

SELECTED BIBLIOGRAPHY

Books

WORKS BY ANNE FRANK

The Diary of a Young Girl. Translated by B. M. Mooyaart-Doubleday. Introduction by Eleanor Roosevelt. Garden City: Doubleday, 1952.

The Diary of Anne Frank: The Critical Edition. Prepared by the Netherlands State Institute for War Documentation. David Barnouw and Gerrold van der Stroom, eds. Translated by Arnold J. Pomerans and B. M. Mooyaart-Doubleday. New York: Doubleday, 1989.

The Diary of Anne Frank: The Revised Critical Edition. Prepared by the Netherlands State Institute for War Documentation. David Barnouw and Gerrold van der Stroom, eds. Translated by Arnold J. Pomerans and B. M. Mooyaart-Doubleday. New York: Doubleday, 2003.

The Diary of a Young Girl: The Definitive Edition. Edited by Otto H. Frank and Mirjam Pressler. Translated by Susan Masotty. New York: Doubleday, 1995.

Anne Frank's Tales from the Secret Annex: Fables, Short Stories, Essays, and an Unfinished Novel by the Author of "The Diary of a Young Girl." Translated by Michel Mok and Ralph Manheim. New York: Washington Square Press, 1983

Aercke, Kristiaan, ed. *Women Writing in Dutch.* New York: Garland Publishing, 1994.

Anne Frank House. *Anne Frank House: A Museum with a Story.* Amsterdam: Anne Frank House, 2001.

Berryman, John. *The Freedom of the Poet.* New York: Farrar, Straus & Giroux, 1976.

Bettleheim, Bruno. *Surviving and Other Essays.* New York: Alfred A. Knopf, 1952.

Bloom, Harold, ed. *A Scholarly Look at* The Diary of Anne Frank. Philadelphia: Chelsea House, 1999.

Brenner, Rachel Feldhay. *Writing as Resistance, Four Women Confronting the Holocaust.* University Park: Pennsylvania State University Press, 1997.

Doneson, Judith E. *The Holocaust in American Film.* Philadelphia: Jewish Publication Society, 1987.

Dresden, Sam. *Persecution, Extermination, Literature.* Translated by Hewy S. Schlogt. Toronto: University of Toronto Press Inc., 1995.

Dwork, Deborah. *Children with a Star: Jewish Youth in Nazi Europe.* New Haven: Yale University Press, 1991.

Enzer, Hyman A., and Sandra Solotaroff-Enzer, eds. *Anne Frank: Reflections on Her Life and Legacy.* Urbana and Chicago: University of Illinois Press, 2000.

Freedom Writers, with Erin Gruwell. *The Freedom Writers Diary.* New York: Broadway Books, 1999.

Gies, Miep, with Alison Leslie Gold. *Anne Frank Remembered: The Story of the Woman Who Helped to Hide the Frank Family.* New York: Simon & Schuster, 1987.

Goodrich, David L. *The Real Nick and Nora.* Carbondale: Southern Illinois University Press, 2001.

Goodrich, Frances, and Albert Hackett. *The Diary of Anne Frank and Related Readings.* Based upon *The Diary of a Young Girl.* Evanston: McDougal Littell, 1997.

Goodrich, Frances, and Albert Hackett. *The Diary of Anne Frank.* Based upon *The Diary of a Young Girl.* Newly adapted by Wendy Kesselman. New York: Dramatists Play Series, Inc., 2000.

Graver, Lawrence. *An Obsession with Anne Frank.* Berkeley: University of California Press, 1995.

Grobman, Alex, and Joel Fishman, eds. *Anne Frank in Historical Perspective: A Teaching Guide for Secondary Schools.* Los Angeles: Martyrs Memorial and Museum of the Holocaust of the Jewish Federation Council, 1995.

Hartman, Geoffrey, ed. *Bitburg in Moral and Political Perspective.* Bloomington: Indiana University Press, 1986.

Hillesum, Etty. *An Interrupted Life: The Diaries of Etty Hillesum.* Translated by Arnold J. Pomerans. New York: Pantheon Books, 1983.

Jones, Judith. *Judith Jones: The Tenth Muse.* New York: Alfred A. Knopf, 2007.

Kopf, Hedda Rosner. *Understanding Anne Frank's "The Diary of a Young Girl." A Student Casebook to Issues, Sources, and Historical Documents.* Westport, Connecticut: Greenwood Press, 1997.

Last, Dick van Galen, and Rolf Wolfswinkel. *Anne Frank and After: Dutch Holocaust Literature in Historical Perspective.* Amsterdam: Amsterdam University Press, 1996.

Lee, Carol Ann. *The Hidden Life of Otto Frank.* New York: Collins Publishers Inc., 2002.

Levi, Primo. *The Drowned and the Saved.* New York: Random House, 1988.

Levin, Meyer. *The Obsession.* New York: Simon & Schuster, 1973.

Lindwer, Willy. *The Last Seven Months of Anne Frank.* Translated by Alison Meersshaert. New York: Pantheon, 1991.

Litvin, Martin. *Audacious Pilgrim: The Story of Meyer Levin.* Woodson, Kansas: Western Books, 1999.

Mak, Geert. *Amsterdam: A Brief Life of the City.* Translated by Philip Blom. London: Harvill Press, 2001.

Melnick, Ralph. *The Stolen Legacy of Anne Frank.* New Haven: Yale University Press, 1997.

Müller, Melissa. *Anne Frank: The Biography.* Translated by Rita and Robert Kimber. New York: Henry Holt, 1998.

Pick, Hella. *Simon Wiesenthal: A Life in Search of Justice.* Boston: Northeastern University Press, 1996.

Presser, Dr. J. *The Destruction of the Dutch Jews.* Translated by Arnold Pomerans. New York: E. P. Dutton & Co., Inc., 1969.

Pressler, Mirjam. *Anne Frank: A Hidden Life.* New York: Puffin Books, 1999.

Robbins, Mari Lu. *A Guide for Using Anne Frank "The Diary of a Young Girl" in the Classroom.* Westminster, CA: Teacher Created Resources, 2007.

Rol, Ruud van der, and Rian Verhoeven, eds. *Anne Frank: Beyond the Diary.* New York: Viking Press, 1993.

Rosenberg, David, ed. *Testimony: Contemporary Writers Make the Holocaust Personal.* New York: Random House, 1989.

Roth, Joseph. *What I Saw.* New York: W. W. Norton & Company, Inc., 2004.

Roth, Philip. *Exit Ghost*. New York: Houghton Mifflin Company, 2007.

Roth, Philip. *The Ghost Writer*. New York: Farrar, Straus & Giroux, 1979.

Schildkraut, Joseph. *My Father and I*. New York: Viking Press, Inc., 1959.

Schloss, Eva, with Julia Kent. *Eva's Story*. New York: St. Martin's Press, 1998.

Schnabel, Ernst. *Anne Frank: A Portrait in Courage*. Translated by Richard Winston and Clara Winston. New York: Harcourt, Brace, 1958.

Sereny, Gitta. *Into That Darkness: An Examination of Conscience*. New York: McGraw-Hill, Inc., 1974.

Shefer-Vanson, Dorothea. *The Diary of Anne Frank,* Cliffs Notes. Hoboken, NJ: Wiley & Sons Publishing, 1984.

Thurman, Judith. *Cleopatra's Nose*. New York: Farrar, Straus & Giroux, 2007.

Velmans, Edith. *Edith's Story*. New York: Soho Press, Inc., 1998.

Ward, C. Geoffrey. *A First-Class Temperament: The Emergence of Franklin Roosevelt*. New York: Harper & Row Publishers, Inc., 1989.

Wiesenthal, Simon. *The Murderers Among Us*. New York: McGraw-Hill, 1967.

Winters, Shelley. *Shelley II: The Middle of My Century*. New York: Simon & Schuster, 1989.

Zweig, Stefan. *The World of Yesterday*. Lincoln: University of Nebraska Press, 1964.

Articles

Alter, Robert. "The View from the Attic: An Obsession with Anne Frank." *New Republic,* December 4, 1995.

Atkinson, Brooks. "Inspired Theater." *New York Times,* October 16, 1955.

Ballif, Algene. "Metamorphosis into American Adolescent." *Commentary,* November 1955. (Reprinted in Enzer, *Anne Frank: Reflections on Her Life and Legacy.*)

Baumel, Judith Tydor. "Teaching the Holocaust through the Diary of Anne Frank." In *Anne Frank in Historical Perspective: A Teaching Guide for Secondary Schools,* Alex Grobman and Joel Fishman, eds. Los Angeles: Martyrs Memorial and Museum of the Holocaust of the Jewish Federation Council, 1995.

Blumenthal, Ralph. "Five Precious Pages Renew Wrangling over Anne Frank." *New York Times,* September 10, 1998.

Brantley, Ben. "This Time, Another Anne Confronts Life in the Attic." *New York Times,* December 5, 1997.

Buruma, Ian. "The Afterlife of Anne Frank." *New York Review of Books,* February 19, 1998.

Canby, Vincent. "A New Anne Frank Still Stuck in the '50s." *New York Times,* December 21, 1997.

Erlenbusch, Sue Jones. "Projects for Anne Frank: Diary of a Young Girl." www.teachervision.com, 1993.

Flanner, Janet. "Letter from Paris." *The New Yorker,* November 11, 1950.

Iskander, Sylvia. "Anne Frank's Reading: A Retrospective." Adapted from "Anne Frank's Reading" in *Children's Literary Association Quarterly* 13, Fall 1988. (Reprinted in Enzer.)

Hackett, Frances. "Diary of the Diary." *New York Times,* September 30, 1956.

Hoagland, Molly Magid. "Anne Frank Onstage and Off," *Commentary,* March 1998.

Johnson, Rebecca Kelch. "Teaching the Holocaust," in *English Journal,* vol. 69, no. 7, October 1980.

Kalb, Bernard. "Diary Footnotes," *New York Times,* October 2, 1955.

Levin, Meyer. "The Child Behind the Secret Door." *New York Times Book Review,* June 15, 1952.

Mulisch, Harry. "Death and the Maiden." *New York Review of Books,* July 17, 1966. (Reprinted in Enzer.)

Nussbaum, Laureen. "Anne Frank." In *Women Writing in Dutch,* Kristiaan Aercke, ed. Garland Publishing, 1994. (Reprinted in Enzer.)

Ozick, Cynthia. "Who Owns Anne Frank?" *New Yorker,* October 6, 1997.

Portman, Natalie. "Thoughts from a Young Actor." *Time,* June 14, 1999.

Probst, Robert. "Literature as Invitation." In *Voices from the Middle,* vol. 8, no. 2, December 2000.

Shore, Lesley. "Anne Frank in Life and Death: Teaching the Lesson of the Holocaust." From Proceedings of the 38th Annual Convention of Jewish Libraries, Toronto, June 15–28, 2003.

Spector, Karen, and Stephanie Jones. "Constructing Anne Frank: Critical Literacy and the Holocaust in Eighth-Grade English." *Journal of Adolescent and Adult Literacy,* September 2007.

Stern, G. B. "Introduction to Tales from the House Behind." Kingswood, England: World's Work, 1952. (Reprinted in Enzer.)

Tynan, Kenneth. "At the Theater: Berlin Postscript," *London Observer,* October 7, 1956.

White, Antonia. Review of *The Diary of a Young Girl,* by Anne Frank. *New Statesman,* May 1953.

Films

Anne Frank Remembered. Directed by Jon Blair. Narrated by Kenneth Branagh. DVD, Sony Picture Classics, 1995.

Anne B. Real. Directed by Lisa France. Performances by Janice Richardson, Carlos Leon, Ernie Hudson. DVD, Screen Media Films, 2003.

The Diary of Anne Frank. Directed by George Stevens. DVD, Twentieth-Century Fox Films, 1959.

PERMISSIONS

INDEX

WORKS AVAILABLE BY
FRANCINE PROSE

READING LIKE A WRITER
A Guide for People Who Love
Books and for Those Who Want to Write Them
ISBN 978-0-06-077705-0 (paperback)

New York Times **Bestseller**

"A love letter about the pleasures of reading and how much writers can learn from the careful reading of great writers as diverse as Virginia Woolf and Flannery O'Connor." —*USA Today*

GOLDENGROVE
A Novel
ISBN 978-0-06-056002-7 (paperback)

"A dazzling mix of directness and metaphor . . . A moving meditation on how, out of the painful passing of innocence and youth, sexuality and identity can miraculously emerge."

—*Los Angeles Times*

A CHANGED MAN
A Novel
ISBN 978-0-06-056003-4 (paperback)

Winner of the Dayton Literary Peace Prize

"Prose uses humor to light up key social issues, to skewer smugness, and to create characters whose flaws only add to their depth and richness." —*Booklist* (starred review)

BLUE ANGEL
A Novel
ISBN 978-0-06-088203-7 (paperback)

National Book Award Finalist in Fiction

"Her trenchant satire of sexual harassment gives political correctness a much deserved poke in the eye." —*Vanity Fair*

THE LIVES OF THE MUSES
Nine Women & the Artists
They Inspired
ISBN 978-0-06-055525-2 (paperback)

"With elegance, eloquence, and majesty, Prose has given us a glimpse of the tangled webs of art and Eros, creativity and inspiration." —*Atlanta Journal-Constitution*

GUIDED TOURS OF HELL
Novellas
ISBN 978-0-06-008085-3 (paperback)

"Wit, knowingness, and an intimate familiarity with guilt and
anxiety—Francine Prose has these qualities in abundance."
—David Lodge, *New York Times Book Review*

THE PEACEABLE KINGDOM
Stories
ISBN 978-0-06-075404-4 (paperback)

"Smartly observed and deftly written, these eleven stories
present the weird jungle of modern life through the eyes of a
wry and mordant writer." —*New York Times Book Review*

PRIMITIVE PEOPLE
A Novel
ISBN 978-0-06-093469-9 (paperback)

"Francine Prose has a wickedly sharp ear for pretentious
American idiom, and no telling detail
escapes her observation."
—*New York Times Book Review*

WOMEN AND CHILDREN FIRST
Stories
ISBN 978-0-06-050728-2 (paperback)

"A meticulously observed collection . . . Stories that glow
with a burnished wisdom." —*New York Times*

HOUSEHOLD SAINTS
A Novel
ISBN 978-0-06-050727-5 (paperback)

"Prose brings off a minor miracle of her own in the rare
sympathy and detachment with which she gives life to this
poignant story. " —Jean Strouse, *Newsweek*

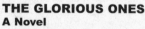

THE GLORIOUS ONES
A Novel
ISBN 978-0-06-149384-3 (paperback)

The Glorious Ones travel through seventeenth-century Italy,
playing commedia dell'arte in the streets and palaces.